Recent contributions to nonlinear partial differential equations

H Berestycki & H Brezis (Editors)
Université Pierre et Marie Curie (Paris VI)

Recent Contributions to Nonlinear Partial Differential Equations

Pitman Advanced Publishing Program
BOSTON · LONDON · MELBOURNE

PITMAN PUBLISHING LIMITED
39 Parker Street, London WC2B 5PB

PITMAN PUBLISHING INC
1020 Plain Street, Marshfield, Massachusetts

Associated Companies
Pitman Publishing Pty Ltd., Melbourne
Pitman Publishing New Zealand Ltd., Wellington
Copp Clark Pitman, Toronto

AMS Subject Classifications: (main) 35-06
(subsidiary) 34-06

Library of Congress Cataloging in Publication Data
Main entry under title:

Berestycki, H. (Henri)
Recent contributions to nonlinear partial differential equations.

(Research notes in mathematics ; 50)
Contributions in English or French.
1. Differential equations, Partial--Addresses, essays, lectures. 3. Differential equations, Nonlinear --Addresses, essays, lectures. I. Brézis, H. (Haim) II. Title. III. Series.
QA374.B347 515.3'53 81-700
ISBN 0-273-08492-5 AACR2

Manufactured in Great Britain

ISBN 0 273 08492 5

Preface

The present volume contains the texts of lectures given at the Université Pierre et Marie Curie (Paris VI) in 1978 and 1979. These lectures were supported by a special Grant from the CNRS (the French National Center for Scientific Research). The aim of this program was to encourage scientific exchange and cooperation between European mathematicians in the domain of nonlinear partial differential equations.

We wish to thank warmly all our colleagues who have participated in this program and who have agreed to write up the text of their lectures. Lastly, we are grateful to Mesdames Chaouche and Force who have worked with much competence on the typing and presentation of this book.

Paris. August 1980

Henri Berestycki - Haïm Brézis

Préface

Le présent ouvrage regroupe les textes de conférences données à l'Université Pierre et Marie Curie (Paris VI) en 1978 et 1979 dans le cadre d'une "ATP Internationale" (Action Thématique Programmée) du Centre National de la Recherche Scientifique. L'objectif de cette action était de promouvoir la coopération entre mathématiciens européens dans le domaine des équations aux dérivées partielles non linéaires.

Nous remercions chaleureusement tous nos collègues qui ont participé à ce programme et qui ont bien voulu rédiger le texte de leurs conférences.
Enfin, nous remercions Mesdames Chaouche et Force qui se sont chargées, avec beaucoup de compétence, de la présentation matérielle de ce volume.

Paris, Aout 1980

Henri Berestycki - Haïm Brézis

Contents

HERBERT AMANN

Saddle points and multiple solutions of nonlinear differential equations

1.

It is the purpose of these talks to report on some results which have recently been obtained by the author in [1] and, jointly with E. Zehnder, in [2]. They concern abstract results on the solvability of equations of the form

$$Au = F(u),$$

in some Hilbert space H, where A is a self-adjoint linear operator and F is a nonlinear potential operator. Since the abstract results are somewhat complicated to state precisely, we begin with indicating some simple applications to the following problems (E), (W), and (H).

Suppose that Ω is a bounded domain in $\mathbb{R}^N$ with smooth boundary, and consider the nonlinear elliptic boundary value problem (BVP)

$$\begin{aligned} -\Delta u &= f(u) \text{ in } \Omega, \\ u &= 0 \quad \text{on } \partial\Omega, \end{aligned} \tag{E}$$

where $f \in C^1(\mathbb{R}, \mathbb{R})$.

The second problem concerns the existence of periodic solutions of nonlinear wave equations. Namely, suppose that $f \in C^2(\mathbb{R}, \mathbb{R})$ satisfies

$$|f'(\xi)| \geq a > 0 \qquad \forall \xi \in \mathbb{R}$$

and some constant $a > 0$. Then we are looking for a solution of the problem

$$\begin{aligned} &u_{tt} - u_{xx} = f(u) \quad \text{in} \quad (0,\pi) \times \mathbb{R}, \\ &u(0,.) = u(\pi,.) = 0 \\ &u(.,t+2\pi) = u(.,t) \quad \forall t \in \mathbb{R}. \end{aligned} \tag{W}$$

Moreover, in each one of the above cases we suppose that

$$f(0) = 0 \text{ and } f'(\infty) := \lim_{|\xi|\to\infty} f'(\xi) \text{ exists,}$$

and by a solution we mean a classical solution.

In the third problem we study the existence of periodic solutions of Hamiltonian systems of ordinary differential equations. More precisely, we suppose that

$$\mathcal{H} \in C^2(\mathbb{R}^{2N}, \mathbb{R}), \quad \mathcal{H}(0) = 0, \quad \mathcal{H}'(0) = 0$$

and

$$\mathcal{H}''(\infty) := \lim_{|z|\to\infty} \mathcal{H}''(z) \text{ exists.}$$

Then, for a given $T > 0$, we are looking for T-periodic (classical) solutions of the system

$$\dot{x} = -\mathcal{H}_y, \quad \dot{y} = \mathcal{H}_x$$

where $z := (x,y) \in \mathbb{R}^N \times \mathbb{R}^N$.

Besides of the above concrete problems we consider the following abstract setting :

H is a real Hilbert space,

A is a self-adjoint linear operator in H,

F is a continuous potential operator on H with potential Φ.

Then we are looking for solutions of the abstract equation

$$Au = F(u). \tag{*}$$

It is easily seen that each one of the problems (E), (W), and (H) is a particular realization of (*). In particular, in the case (H) the linear operator A is defined by

$$\mathrm{dom}(A) := \{z = (x,y) \in H^1(0,T;\mathbb{R}^{2N}) \mid z(0) = z(T)\}$$

$$Az := \{\dot{y}, -\dot{x}\} .$$

For the following it will be important to know the structure of the spectrum of the operator A. It is either well-known or easily seen that $\sigma(A)$ is a pure point spectrum in each case. More precisely, in the elliptic case (E), $\sigma(A)$ consists of an increasing sequence of eigenvalues

$$\lambda_o < \lambda_1 < \lambda_2 < \cdots$$

of finite multiplicity $m(\lambda_j)$, such that $\lambda_j \to \infty$ as $j \to \infty$.

In the case (W), $\sigma(A)$ is given by the set $\{j^2 - k^2 \mid (j,k) \in \mathbb{N}^* \times \mathbb{Z}\}$. Hence if we order this set in the natural way such that

$$\cdots < \lambda_{-2} < \lambda_{-1} < \lambda_o = 0 < \lambda_1 < \lambda_2 \cdots ,$$

it follows that $\lambda_j \to \pm\infty$ as $j \to \pm\infty$, respectively, each λ_j, $j \in \mathbb{Z} \setminus \{0\}$, is an eigenvalue of finite multiplicity, and $\lambda_o = 0$ is an eigenvalue of infinite multiplicity.

In the case of problem (H) it is not difficult to see that

$$\sigma(A) = \sigma_p(A) = \tau\mathbb{Z}, \text{ where } \tau := 2\pi/T,$$

and each $\lambda_j \in \sigma(A)$ is an eigenvalue of multiplicity 2N.

Since we have assumed that $f(0) = 0$ or $\mathcal{H}'(0) = 0$, respectively, each one of the problems (E), (W), or (H) has the trivial solution $u = 0$. Hence we are interested in the existence of nontrivial solutions.

Observe that the existence of $f'(\infty)$ or $\mathcal{H}''(\infty)$, respectively, implies that f or $\mathcal{H}'$, resp., is asymptotically linear (which we assume here for simplicity). Under hypotheses of this type much work has already been done for the elliptic case, for example by Ambrosetti, Mancini, Rabinowitz, Castro-Lazer, and very recently by Coron [3] and Thews [4] , to name only a few.

For example Coron [3] obtains the existence of a nontrivial solution if

$$f'(\xi) \leq \lambda_{k+1} \quad \text{and} \quad f'(0) < \lambda_k < f'(\infty)$$

for some $k \in \mathbb{N}$ and all $\xi \in \mathbb{R}$.

Thews [4] obtains essentially the same result. In addition he obtains also the existence of a nontrivial solution in the "dual" case, that is, if

$$f'(\xi) \geq \lambda_{k-1} \quad \text{and} \; f'(\infty) < \lambda_k < f'(0).$$

Much less seems to be known for the wave equation (W). Brezis and Nirenberg [5] obtain the existence of a nontrivial solution of (W) if

$$f' \geq 0, \; f'(\infty) < \lambda_1 < f'(0)$$

and if the number of squares j^2, $j \in \mathbb{N}$, less than $f'(0)$, is odd.

This result has been generalized (in some sense) by Mancini [6] , who obtains the existence of a nontrivial solution of (W) if either

$$f' \geq \lambda_{k-1} \geq 0, \; f'(\infty) < \lambda_k < f'(0) \tag{1}$$

or

$$f' \leq \lambda_{-k+1} \leq 0, \; f'(0) < \lambda_{-k} < f'(\infty) \tag{2}$$

for some $k \in \mathbb{N}$, and $f'(0)$, $f'(\infty) \notin \sigma(A)$.

Finally, Coron [3] obtains the existence of a nontrivial solution of (W) if

$$f' \leq 0 \quad \text{and} \quad f'(0) < \lambda_{-1} < f'(\infty) < 0$$

(In all of the above results we have not given the most general regularity and growth conditions, but we have pointed out the basic assumptions concerning the relations between $f'(0)$, $f'(\infty)$, and $\sigma(A)$.)

The crucial point in each one of the above results is the fact that the corresponding problem (E) or (W) has a nontrivial solution if there is at least one eigenvalue of A between $f'(0)$ and $f'(\infty)$. However this fact has only been shown under additional hypotheses (be it that $f' \leq \lambda_{k+1}$, $f' \geq \lambda_{k-1}$, or that $f'(0) < f'(\infty)$ if $\lambda_k \geq 0$ as in (1), or $f'(\infty) < f'(0)$ if $\lambda_k \leq 0$ as in (2)).

The following theorem which is a very special case of our general abstract results shows that none of these restrictions is necessary, provided we exclude resonance at zero and infinity.

<u>THEOREM (E), (W)</u> : Consider problem (E) or (W), respectively, and suppose that $f'(0)$, $f'(\infty) \notin \sigma(A)$. Then there exists at least one nontrivial solution if there is at least one eigenvalue of A between $f'(0)$ and $f'(\infty)$.

As an application of our abstract results to problem (H) we present the following.

THEOREM (H) : Suppose that there exists an $\alpha > 0$ such that $\sigma(\mathcal{H}''(z)) \subset [\alpha,\infty)$ for all $z \in \mathbb{R}^N$. Moreover, suppose that

$$(\mathcal{H}''(0)z,z) = \sum_{k=1}^{N} \alpha_k^0 (x_k^2 + y_k^2)$$

$$(\mathcal{H}''(\infty)z,z) = \sum_{k=1}^{N} \alpha_k^\infty (x_k^2 + y_k^2)$$

for all $z = (x,y) \in \mathbb{R}^N \times \mathbb{R}^N$. Then, for every $\tau > 0$ let

$$i(\tau) := \sum_{k=1}^{N} \Big(\sum_{\alpha_k^0 < \tau j \le \alpha_k^\infty} 1 \; - \sum_{\alpha_k^\infty < \tau j \le \alpha_k^0} 1 \Big),$$

where j runs through $\mathbb{Z}$ in this summation.

Then, for every $\tau > 0$ for which

$$i(\tau) \neq 0 \text{ and } \alpha_k^0, \alpha_k^\infty \notin \tau\mathbb{Z} \quad k = 1,\dots,N,$$

there exists a nonconstant $2\pi/\tau$-periodic solution of (H).

To compute the index $i(\tau)$ for a given $\tau > 0$ such that $\alpha_k^0, \alpha_k^\infty \notin \tau\mathbb{Z}$, $k = 1,\dots,N$, one has, for each $k \in \{1,\dots,N\}$, to count the number of points in $\tau\mathbb{Z}$ between α_k^0 and α_k^∞. This number has to be given a + sign if $\alpha_k^0 < \alpha_k^\infty$ and a - sign if $\alpha_k^0 > \alpha_k^\infty$. Then the total (algebraic) sum of these numbers equals $i(\tau)$. For example, in the following example (N = 3)

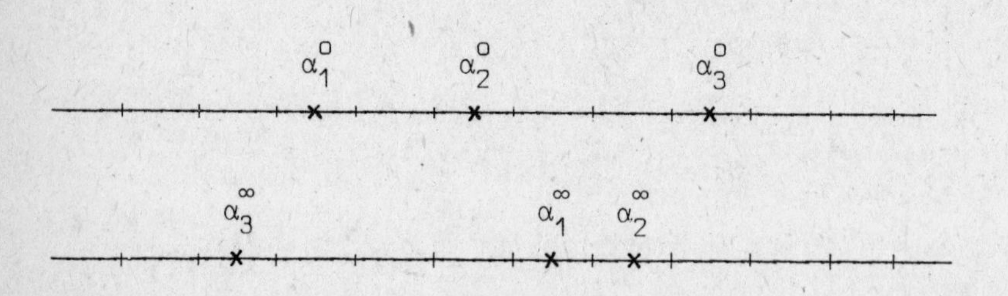

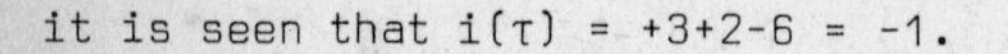
it is seen that $i(\tau) = +3+2-6 = -1$.

We want to point out that in the above Theorem (H) the spectra of the linearizations $\mathcal{H}''(0)$ and $\mathcal{H}''(\infty)$, resp., of $\mathcal{H}'$ at zero and infinity, resp., are allowed to overlap arbitrarily. This is in sharp contrast to existence results (for Hamiltonians having at most quadratic growth, as in our case) obtained recently by Clarke and Ekeland [7] by completely different methods. In that paperit is assumed that $\mathcal{H}$ is convex and that there exist constants $k < K$ such that $\sigma(\mathcal{H}''(\infty)) \subset [0,k]$ and $\sigma(\mathcal{H}''(0)) \subset [K,\infty]$, that is, these spectra are strictly separated. (Again we indicate only the crucial point of the hypotheses. In [7] $\mathcal{H}$ is not even supposed to be differentiable). Moreover, we should like to point out that our results apply also to nonconvex Hamiltonians (cf.[2]), whereas convexity of $\mathcal{H}$ is essential for the techniques used by Clarke and Ekeland.

2.

In this section we consider the general problem (P), and we impose the following additional <u>hypothesis (H)</u> :

There exist real numbers $\alpha < \beta$ belonging to the resolvent set $\rho(A)$ of A such that $\sigma(A) \cap (\alpha,\beta)$ consists of at most finitely many eigenvalues of finite multiplicity. Moreover, (H)

$$\alpha \|u-v\|^2 \leq \langle F(u)-F(v),u-v\rangle \leq \beta\|u-v\|^2$$

for all $u,v \in H$.

We emphasize the fact that no assumption whatsoever concerning the spectrum of A outside $[\alpha,\beta]$ has been made.

Formally the equation $Au = F(u)$ is the Euler equation of the variational problem

$$\phi(u) := -\frac{1}{2} < Au,u > + \Phi(u) \Rightarrow \text{extr.}$$

However the functional ϕ is only defined on the domain of A and, hence, is not C^1. For this reason we remplace first ϕ by another function which is continuously differentiable and whose set of critical points is equivalent to the set of all solutions of $Au = F(u)$.

We decompose H into an orthogonal sum

$$H = X \oplus Y \oplus Z,$$

where X, Y, and Z are invariant subspaces of A such that

$$\sigma(A|X) = \sigma(A) \cap (-\infty,\alpha), \ \sigma(A|Y) = \sigma(A) \cap (\beta,\infty), \text{ and } \sigma(A|Z) = \sigma(A) \cap (\alpha,\beta).$$

Then, by means of the spectral representation of A we define symmetric bounded linear operators $R \in \mathcal{L}(H,X)$, $S \in \mathcal{L}(H,Y)$, and $T \in \mathcal{L}(H,Z)$ such that

$$-R^2 + S^2 + T^2 = (A-\alpha)^{-1}.$$

Finally, we let

$$\Gamma(u) := \Phi(u) - \frac{\alpha}{2}\|u\|^2 \qquad \forall u \in H,$$

and we define $f : X \times Y \times Z \to \mathbb{R}$ by

$$f(x,y,z) := \frac{1}{2}(\|x\|^2 - \|y\|^2 - \|z\|^2) + \Gamma(Rx+Sy+Tz).$$

Using the above definitions it is not difficult to verify the validity of the following

LEMMA 1: $f \in C^1(X \times Y \times Z, \mathbb{R})$ and $f'(x,y,z) = 0$ iff $Rx + Sy + Tz$ is a solution of $Au = F(u)$.

Next, using the second part of hypothesis (H), one verifies that

$$\forall (y,z) \in Y \times Z : f(.,y,z) : X \to \mathbb{R} \text{ is strongly convex,}$$

$$\forall (x,z) \in X \times Z : f(x,.,z) : Y \to \mathbb{R} \text{ is strongly concave.}$$

This implies that, for every $z \in Z$, the function

$$f(.,.,z) : X \times Y \to \mathbb{R}$$

possesses a unique saddle point $(x(z),y(z))$, from which it follows, in particular, that

$$D_1 f(x(z),y(z),z) = 0$$

$$D_2 f(x(z),y(z),z) = 0$$

for every $z \in Z$.

Now we let

$$g(z) := f(x(z),y(z),z) \quad \forall z \in Z.$$

By using the particular properties of saddle points, it can be shown that the following lemma is true.

LEMMA 2 : $g \in C^1(Z,\mathbb{R})$ and $g'(z) = 0$ iff $Rx(z) + Sy(z) + Tz$ is a solution of $Au = F(u)$.

Observe that we have not excluded the fact that $Z = \{0\}$, which occurs precisely if $\sigma(A) \cap (\alpha,\beta) = \emptyset$. Hence we obtain as a simple application of the above results the following existence and uniqueness theorem.

THEOREM 1 : Suppose that $\sigma(A) \cap (\alpha,\beta) = \emptyset$.
Then the equation $Au = F(u)$ is uniquely solvable.

Due to Lemma 2 we have reduced our original problem to the problem of finding critical points of the functional g on the finite-dimensional space Z. In order to make further progress we impose now the following hypothesis (F_∞) concerning the asymptotic behavior of F.

$$\left.\begin{array}{l}\text{There exists a symmetric linear operator}\\ F'(\infty) \in \mathcal{L}(H)\text{, which commutes with A and}\\ \text{satisfies } \sigma(A) \cap \sigma(F'(\infty)) = \emptyset\text{, such}\\ \text{that}\\ \lim\limits_{\|u\| \to \infty} \dfrac{F(u)-F'(\infty)u}{\|u\|} = 0.\end{array}\right\} \qquad (F_\infty)$$

Clearly the assumption that $\sigma(A) \cap \sigma(F'(\infty)) = \emptyset$ is a "nonresonance condition at infinity", whereas the commutativity hypothesis is of technical nature, allowing sharp estimates for g.

As a consequence of (F_∞) we obtain now the following

LEMMA 3 : Let (F_∞) be true. Then there exist constants $\gamma > 0$ and $\delta \geq 0$ such that

$$\|g'(z)\| \geq \gamma \|z\| - \delta \quad \forall z \in Z.$$

Recall that we want to find critical points of g. For this purpose we consider the gradient flow on Z defined by

$$\dot{z} = -g'(z), \tag{3}$$

and we denote by S the set of bounded solutions of (3). If $S \neq \emptyset$, then it is an easy consequence of the fact that (3) is a gradient flow, that there exists at least one rest point, that is, at least one critical point of g.

(It should be remarked that hypothesis (H) implies that g' is linearly bounded and Lipschitz continuous. Hence the flow (3) is well defined.)

It is an easy consequence of Lemma 3 that S is compact. Hence it is an isolated invariant set and possesses a generalized Morse index h(S) in the sense of C.C. Conley [8] . This Morse index is a homotopy type of pointed topological spaces. Using the full strength of condition (F_∞), we are able to compute h(S) (via an appropriate continuation argument), and we find that

$$h(S) = [\textstyle\sum^{m_\infty}],$$

the homotopy type of a pointed m -sphere, where m_∞ is the index of the symmetric linear operator $[A-F'(\infty)] \mid Z \in \mathcal{L}(Z)$, that is, the dimension of a maximal linear subspace of Z on which $A-F'(\infty)$ is negative definite. This implies in particular that $h(S) \neq h(\emptyset)$, hence $S \neq \emptyset$. Thus g has at least one critical point, which implies the following

THEOREM 2 : Let (F_∞) be satisfied. Then the equation $Au = F(u)$ has at least one solution.

Now we suppose that $F(0) = 0$. Hence $Au = F(u)$ has the trivial solution $u = 0$, and we are interested in the existence of nontrivial solutions. For this purpose we impose the following hypothesis (F_0).

$$\left.\begin{array}{l} F(0) = 0,\ F \text{ is Gateaux-differentiable at } 0 \text{ with} \\ \text{a symmetric derivative } F'(0) \in \mathcal{L}(H), \text{ which commutes} \\ \text{with } A \text{ and satisfies } \sigma(A) \cap \sigma(F'(0)) = \emptyset. \end{array}\right\} \quad (F_0)$$

In addition we have to impose some further regularity conditions which are too complicated to be given here. However these regularity conditions are satisfied in all of our applications. In particular they are satisfied if we suppose that F is (Fréchet-)differentiable at 0, which is, however, too strong an assumption for being met in concrete cases.

Using condition (F_0), it can be shown that $S_0 := \{0\}$ is an isolated invariant set for the flow (3). Moreover, similarly as above, we can compute the homotopy index $h(S_0)$ of S_0. In this case we find that

$$h(S_0) = [\textstyle\sum^{m_0}],$$

where m_0 is the index of $[A-F'(0)]\,|\,Z \in \mathcal{L}(Z)$.
If we assume now that $m_0 \neq m_\infty$, then it follows that $h(S_0) \neq h(S)$, which implies, in particular, that $S \neq S_0$. Consequently, there must exist a nonempty compact isolated invariant set of (3) in $Z \setminus \{0\}$. Thus, (3) being a gradient flow, the existence of a nonzero critical point of g follows. By this way we prove the following.

THEOREM 3 : Let hypotheses (F_0) and (F_∞) (and an appropriate regularity condition) be satisfied, and assume that $m_0 \neq m_\infty$. Then the equation $Au = F(u)$ has at least one nontrivial solution.

In concrete applications of this theorem one has to compute the numbers m_0 and m_∞. This can often be done, for example in the problems (E), (W), and (H) considered above. By this way we can deduce Theorem (E), (W) and Theorem (H) from the general Theorem (3). For details we refer to [2] .

Bibliography

1 H. Amann, Saddle points and multiple solutions of differential equations. Math. Zeitschr., 169 (1979), 127-166

2 H. Amann and E. Zehnder, Multiple solutions for a class of nonresonance problems and applications to differential equations. to appear in Annali Scuola Normale Sup. Pisa.

3 J. M. Coron, Résolution de l'équation Au + Bu = f où A est linéaire autoadjoint et B déduit d'un potentiel convex. Preprint 1978.

4 K. Thews, A reduction method for some Dirichlet problems. J. Nonlinear Analysis, Theory, Methods & Appl., 3 (1979), 795-813

5 H. Brezis and L. Nirenberg, Forced vibrations for a nonlinear wave equation. C.P.A.M. XXXI (1978), 1-30.

6 G. Mancini, Periodic solutions of some semilinear autonomous wave equations. Boll. U.M.I. (5), 15-B (1978), 649-673.

7 F.H. Clarke and I. Ekeland, Hamiltonian trajectories having prescribed minimal period. Preprint 1978.

8 C.C. Conley, Isolated invariant sets and the Morse index. CBMS Regional Conference Series in Math., 38 (1978), AMS, Providence, R.I..

Herbert AMANN
Mathematisches Institut
Universität Zurich
Freiestrasse 36
8032 ZURICH
SWITZERLAND

HERBERT AMANN

Multiple periodic solutions of autonomous Hamiltonian systems

1.

In this talk we describe some existence theorems which we obtained recently, jointly with E. ZEHNDER [1].

Throughout we suppose that

$$\mathcal{H} \in C^2(\mathbb{R}^{2N}, \mathbb{R}),\ \mathcal{H}(0) = 0,\ \mathcal{H}'(0) = 0,$$

$$\mathcal{H} \text{ is even, and } \mathcal{H}''(\infty) := \lim_{|z| \to \infty} \mathcal{H}''(z) \text{ exists.}$$

Then we are looking for nontrivial, T-periodic solutions of the Hamiltonian system

$$\dot{x} = -\mathcal{H}_y,\ \dot{y} = \mathcal{H}_x, \qquad (*)$$

where $z = (x,y) \in \mathbb{R}^N \times \mathbb{R}^N$, and $T > 0$ is given.

In the following we let

$$\mathcal{H}_0(z) := \frac{1}{2}(\mathcal{H}''(0)z,z)$$
$$\mathcal{H}_\infty(z) := \frac{1}{2}(\mathcal{H}''(\infty)z,z)$$

for all $z \in \mathbb{R}^{2N}$, that is $\mathcal{H}_0(z)$ (or $\mathcal{H}_\infty(z)$, resp.) is the quadratic term of the Taylor expansion of $\mathcal{H}$ at 0 (or at ∞, resp.). Then we claim the validity of the following

THEOREM 1 : Suppose that

$$\mathcal{H}_0(z) = \frac{1}{2} \sum_{k=1}^{N} \alpha_k^0 (x_k^2 + y_k^2)$$

and

$$\mathcal{H}_\infty(z) = \frac{1}{2} \sum_{k=1}^{N} \alpha_k^\infty (x_k^2 + y_k^2)$$

for all $z \in \mathbb{R}^{2N}$ and some $\alpha_1^0, \ldots, \alpha_N^0, \alpha_1^\infty, \ldots, \alpha_N^\infty \in \mathbb{R}$.

For every $\tau > 0$, let

$$i(\tau) := \sum_{k=1}^{N} \Big(\sum_{\alpha_k^0 < \tau j \leq \alpha_k^\infty} 1 - \sum_{\alpha_k^\infty < \tau j \leq \alpha_k^0} 1 \Big),$$

where j runs through $\mathbb{Z}$ in this summation.

Then, for every $\tau > 0$ such that $\alpha_k^0, \alpha_k^\infty \notin \tau\mathbb{Z}$, $k = 1, \ldots, N$, there exist at least $|i(\tau)|$ nonzero pairs $(z,-z)$ of geometrically distinct $\frac{2\pi}{\tau}$ -periodic solutions of $(*)$.

REMARKS : (1) To compute the index $i(\tau)$ for a given $\tau > 0$ such that $\alpha_k^0, \alpha_k^\infty \notin \tau\mathbb{Z}$, $k = 1, \ldots, N$, one has, for each $k \in \{1, \ldots, N\}$, to count the number of points in $\tau\mathbb{Z}$ between α_k^0 and α_k^∞. This number has to be given a + sign if $\alpha_k^0 < \alpha_k^\infty$ and a - sign if $\alpha_k^\infty < \alpha_k^0$. Then the total (algebraic) sum of these numbers equals $i(\tau)$. For example, in the following case $(N = 6)$

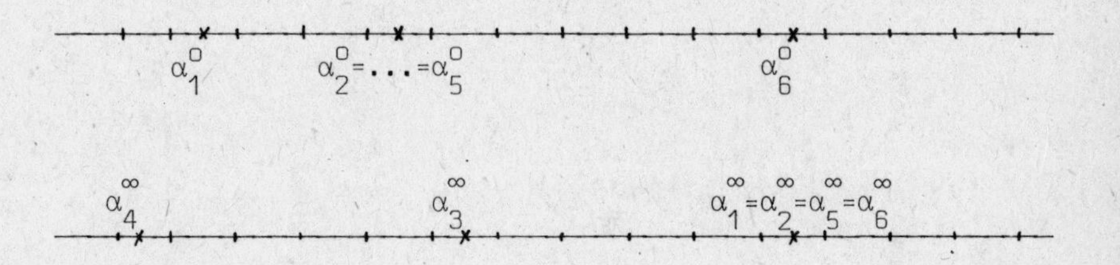

it is verified that $i(\tau) = 9+6+1-4+6+0 = 18$.

(2) In general $|i(\tau)| \to \infty$ as $\tau \to 0$. Thus Theorem 1 guarantees, in general, the existence of very many periodic solutions having a very large period.

(3) In the above theorem nonzero critical points of the Hamiltonian are considered as (constant) periodic solutions. However, if $\mathcal{H}$ is strictly convex, them all nonzero solutions of (*) are nonconstant.

(4) There are more general versions of Theroem 1. In particular, $\alpha_i^o \in \tau\mathbb{Z}$ is allowed. But then the statement of the assertion is more complicated (cf. [1]).

(5) It seems worthwhile to emphasize the fact that there are no sign restrictions of the "normal modes" α_k^o, α_k^∞, $k=1,\ldots,N$.

In the following theorem we consider a situation where the Birkhoff normal forms of the linearized vector fields at 0 and ∞ are of distinct type.

THEOREM 2 : Suppose that

$$\mathcal{H}_o(z) = \frac{1}{2} \sum_{k=1}^{N} \alpha_k^o(x_k^2 + y_k^2)$$

and

$$\mathcal{H}_\infty(z) = \frac{1}{2} \sum_{k=1}^{N} \alpha_k^\infty(x_k^2 - y_k^2)$$

for all $z \in \mathbb{R}^{2N}$. Morever, suppose that $\alpha_k^\infty \neq 0$ for all $k = 1,\ldots,N$, and

$$\alpha^o := \min \{\alpha_k^o\} > \alpha^\infty := \max \{|\alpha_k^\infty|\}.$$

Then, for every $\tau > 0$, there exist at least $Nm(\tau)$ nonzero pairs $(z,-z)$ of $\frac{2\pi}{\tau}$-periodic solutions of (*), where

$$m(\tau) := \text{card } [(\alpha^\infty, \alpha^o) \cap \tau\mathbb{Z}].$$

Since $m(\tau)$ equals the number of lattice points (of the lattice $\tau\mathbb{Z}$) in the open interval $(\alpha^{\infty},\alpha^{0})$, it is obvious that $m(\tau) \to \infty$ as $\tau \to 0$. Thus Theorem 2 guarantees again the existence of very many nontrivial periodic solutions of (*) of very large period.

It should be remarked that the above theorems (which are special cases of much more general results given in [1]) are the only <u>global</u> multiplicity results known to the authors concerning the existence of periodic solutions with a <u>fixed</u> period. There seems to be only one global multiplicity result, namely the recent result of Ekeland and Lasry [2], which guarantees the existence of N periodic solutions of (*) having prescribed energy but unknown period. Needless to say that the methods of these two papers are completely different.

2.

The above theorems are deduced by means of Lusternik-Schnirelman theory arguments. Since their proofs are rather involved we shall here indicate only the basic ideas.

<u>STEP 1</u> : We consider the Hamiltonian system (*) as a concrete realization of a nonlinear equation of the form

$$Au = F(u)$$

in the Hilbert space $H := L_2(0,T;\mathbb{R}^{2N})$. Here A is defined by

$$\operatorname{dom}(A) := \{z \in H^1(0,T;\mathbb{R}^{2N}) \mid z(0) = z(T)\}$$

and

$$A\{x,y\} = \{\dot{y},-\dot{x}\} \qquad \forall z = (x,y) \in \operatorname{dom}(A).$$

It is not difficult to see that A is self-adjoint and that

$\sigma(A) = \sigma_p(A) = \tau\mathbb{Z}$, where $\tau := 2\pi/T$, and each $\lambda \in \sigma(A)$ is an eigenvalue of multiplicity $m(\lambda) = 2N$.

The nonlinear map F is defined by

$$F(z)(t) := \mathcal{H}'(z(t)) \qquad \forall t \in (0,T)$$

and all $z \in H$. Hence F is a continuous odd potential operator on H, and there exist numbers $\alpha,\beta \in \mathbb{R}$ such that

$$\alpha\|u-v\|^2 \leq \langle F(u)-F(v),u-v\rangle \leq \beta\|u-v\|^2 \qquad \forall u,v \in H.$$

Finally, every T-periodic solution of (*) corresponds to a solution of $Au = F(u)$, and vice versa.

STEP 2 : Using the abstract properties of A and F and a reduction method introduced by the author in [3], it can be shown that there exists a finite-dimensional subspace Z of H and an even function $g \in C^1(Z,\mathbb{R})$ satisfying $g(0) = 0$ and the Palais-Smale condition such that the set of critical points of g is in a one-to-one correspondence with the set of the solutions of the equation $Au = F(u)$.

STEP 3 : In order to find critical points of g we employ a variant of the Lusternik-Schnirelman theory, due to D. Clark [4]. For this purpose we recall that the genus gen(B) of a closed symmetric subset B of Z such that $0 \notin B$ is defined to be the least integer k such that there exists an odd continuous map of B in $\mathbb{R}^k\setminus\{0\}$. For example, as a consequence of the Borsuk-Ulam theorem it follows that $\mathrm{gen}(S^k) = k + 1$, where S^k is the k-sphere.

It is now standard to define

$$c_k(g) := \inf_{\text{gen}(B)\geq k} \sup_B (g)$$

and it is easily seen that

$$-\infty \leq c_1(g) \leq \ldots \leq c_{\dim(Z)}(g) \leq 0.$$

For every $c \in \mathbb{R}$ let

$$K_c := \{z \in Z \mid g(z) = c,\ g'(z) = 0\}$$

and suppose that, for some $\ell, m \in \mathbb{N}^*$,

$$-\infty < c_{\ell+1}(g) = \ldots = c_{\ell+m}(g) =: c < 0.$$

Then it has been shown by Clark [4] that

$$\text{gen}(K_c) \geq m.$$

Now, given the hypotheses of the above theorems, by means of precise estimates of the functional g it is possible to show that there exist numbers $\ell, M \in \mathbb{N}^*$ such that M is even and

$$-\infty < c_{\ell+r}(g) < 0 \qquad r \in \{1,\ldots,M\}. \tag{**}$$

Hence Clark's result implies the existence of at least M **pairs** $(z,-z)$ of nonzero critical points of g. Moreover, if

$$c_{s+1}(g) = c_{s+2}(g) = \ldots = c_{s+k}(g)$$

for some $s, k \in \mathbb{N}^*$ for which $c_{s+1}(g) \in (-\infty, 0)$, it follows that

$$\text{gen}(K_{c_{s+1}}) \geq k. \tag{***}$$

STEP 4 : It follows from steps 1 and 2 that every critical point of g corresponds to a T-periodic solution of (*). However, suppose that u is a solution of (*) and let $s \in \mathbb{R}$ be given. Then it is obvious that the function

$$u_s(t) := u(t+s) \qquad \forall t \in \mathbb{R}$$

represents the same periodic orbit of (*), although different parameters s will, in general, define distinct points in the function space H. In other words, the equation $Au = F(u)$ has many more solutions than there are geometrically distinct periodic solutions of (*). Hence we have to eliminate the artificial multiplicity which is contained in the abstract formulation in order to get an estimate on the number of geometrically distinct periodic solutions of (*).

The key observation for this reduction is the following simple

LEMMA 1 : Suppose that u is a solution of (*) and $\phi \in C^1(\mathbb{R},\mathbb{R})$ is a reparametrization such that $u \circ \phi$ is also a solution of (*). Then there exists an $s \in \mathbb{R}$ such that $\phi(t) = t + s$ for all $t \in \mathbb{R}$.

STEP 5 : Due to the T-periodicity and to Lemma 1 we define a strongly continuous unitary representation of the circle group S^1,

$$S^1 \to (H), \quad \sigma \mapsto U_\sigma ,$$

where $\sigma = e^{i\tau s} \in S^1$ and

$$U_\sigma u(t) = u(t+s).$$

Then it is easily seen that U_σ commutes with A and with F. Moreover, since Z is a reducing subspace of A, it follows that this representation defines a continuous (but not free!) action of S^1 on Z. Next we establish the following

LEMMA 2 : g is invariant with respect to the action of S^1 on Z. More precisely,

$$g \circ U_\sigma = g \qquad \forall \sigma \in S^1.$$

In the following we let

$$\mathcal{O}(z) := \{U_\sigma z \mid \sigma \in S^1\}$$

be the orbit (of this action) through the point $z \in Z$. As a consequence of the above considerations and the evenness of g we obtain then

LEMMA 3 : For every $z \in K_c(g)$,

$$\mathcal{O}(z) \cup \mathcal{O}(-z) \subset K_c(g).$$

Distinct orbits of critical points of g correspond to geometrically distinct T-periodic solutions of (*).

STEP 6 : It follows from Lemma 3 that one has to estimate the number of critical orbits of g in order to get an estimate for the number of geometrically distinct T-periodic solutions of (*). To do this we consider, for any given $z \neq 0$, the isotropy subgroup S^1_z of z, that is,

$$S^1_z := \{\sigma \in S^1 \mid U_\sigma z = z\}.$$

Since S^1_z is a closed subgroup of S^1, it follows that either $S^1_z = S^1$ or S^1_z is a cyclic subgroup of S^1 of some finite order, say m. In the first case $\mathcal{O}(z) = \{z\}$ and gen $(\mathcal{O}(z) \cup \mathcal{O}(-z)) = 1$.

In the second case the map

$$S^1 \to Z,\ \sigma \to U_{\sigma^{1/m}}(z)$$

is a homomorphism of S^1 onto $\mathcal{O}(z)$. If $\mathcal{O}(z) = \mathcal{O}(-z)$, it is easily seen that this homeomorphism is odd. Hence

$$\text{gen}(\mathcal{O}(z) \cup \mathcal{O}(-z)) = \text{gen}(\mathcal{O}(z)) = \text{gen}(S^1) = 2.$$

Otherwise $\mathcal{O}(z) \cap \mathcal{O}(-z) = \emptyset$. But then it is easy to define an odd continuous map

$$\mathcal{O}(z) \cup \mathcal{O}(-z) \to S^1,$$

simply by mapping $-U\ (z) = U_\sigma(-z)$ onto $-\sigma \in S^1 \subset \mathbb{R}^2$ if $U_\sigma(z)$ is being mapped onto σ by the above homeomorphism. Thus in each case we obtain the following result :

LEMMA 4 : For every $z \in Z \setminus \{0\}$,

$$\text{gen}(\mathcal{O}(z) \cup \mathcal{O}(-z) \leq 2.$$

STEP 7 : Coming back to formula (**), we see from Lemma 3 that we obtain M nonzero pairs of geometrically distinct T-periodic solutions of (*) if the M levels $c_{\ell+r}(g)$ are pairwise distinct. If two of them coincide, say

$$c_{s+1}(g) = c_{s+2}(g),$$

then $\text{gen}(K_{c_{s+1}}) \geq 2$ by (***). In this case all critical points on the level c_{s+1} may lie on the orbits $\mathcal{O}(z) \cup \mathcal{O}(-z)$ for some $z \in Z \setminus \{0\}$. In this case it follows from Lemma 4 that we obtain at least M/2 nonzero pairs (z,-z) of geometrically distinct T-periodic solutions of (*). If, finally, at least 3 critical levels coincide, i.e.,

$$c_{s+1}(g) = c_{s+2}(g) = c_{s+3}(g),$$

then it follows from the properties of the genus, Lemma 4, and the estimate (***), that there are infinitely many distinct orbits $\sigma(z)$ in the critical set $K_{c_{s+1}}$. Consequently, we see that the number of nonzero pairs $(z,-z)$ of geometrically distinct T-periodic solutions of (*) is at least M/2.

Finally the assertions of the above theorems are obtained by evaluating the conditions which lead to (**). For details we refer to [1].

Bibliography

1 H. Amann and E. Zehnder, Periodic solutions of asymptotically linear Hamiltonian systems. To appear.

2 I. Ekeland and J.M. Lasry, Nombre de solutions périodiques des equations de Hamilton. Preprint 1978.

3 H. Amann, Saddle points and multiple solutions of differential equations. Math. Zeitschr., 169 (1979), 127-166.

4 D. Clark, A variant of the Lusternik-Schnirelman theory. Indiana Univ. Math. J., 22 (1972), 65-74.

Herbert AMANN

Mathematisches Institut
Universität Zurich
Freiestrasse 36
CH-8032 ZURICH
SWITZERLAND

ANTONIO AMBROSETTI & GIOVANNI MANCINI

On some free boundary problems

0. INTRODUCTION

This paper deals with two free boundary problems : (i) the study of vortex rings appearing in an ideal fluid and (ii) the equilibrium of a plasma confined in a toroidal cavity (the Tokamak machine).

The results concerning problem (ii) improve several previous papers (Berestycki and Brézis, 1976; Puel, 1977; Temam, 1977) and will appear in a more complete form in (Ambrosetti and Mancini, to appear). As for problem (i), we follow closely a work by Fraenkel and Berger (Fraenkel and Berger, 1974) showing how suitable modifications of their arguments allow to cover the case in which the vortex-strength parameter is a priori given.

During the preparation of this part, we learned that results concerning geometric properties of solutions which are important in dealing with several free boundary problems as (i-ii) in particular, seem to have been obtained by several people : e.g. the paper (Bona, Bose and Turner, To appear) on water waves, where such informations are obtained under weaker assumptions and, more generally, the paper (Gidas, Ni and Nirenberg, To appear).

1. VORTEX RINGS IN AN IDEAL FLUID

(A) The result by Fraenkel and Berger.

Let $\Pi = \{(r,z) \in \mathbb{R}^2 : r > 0\}$ and $L = r\frac{\partial}{\partial r}(\frac{1}{r}\frac{\partial}{\partial r}) + \frac{\partial^2}{\partial z^2}$. The existence of steady vortex rings in an ideal fluid, whose velocity is assumed axisymmetric is described by the following free boundary problem (Fraenkel and

Berger, 1974) :

Given $\lambda > 0$, $k > 0$, $W > 0$ and $f : \mathbb{R}^+ \to \mathbb{R}^+$, to find an open subset $A \subset \Pi$ and a C^1 function $\psi(r,z)$ such that

$$- L\,\Psi = \lambda r^2 f(\Psi) \quad \text{in } A \tag{1a}$$

$$L\,\Psi = 0 \quad \text{in } \Pi \setminus A \tag{1b}$$

$$\Psi > 0 \quad \text{in } A \text{ and } \Psi = 0 \text{ on } \partial A \tag{1c}$$

$$\Psi(0,z) = -k \tag{1d}$$

$$\frac{1}{r}\Psi_r \to -W \text{ and } \frac{1}{r}\Psi_z \to 0 \text{ as } r^2+z^2 \to \infty \tag{1e}$$

Here, W represent the velocity of the fluid at infinity, and f, the so-called vorticity function, satisfies

$$f \in C^2,\ f(0) = 0,\ f(s) > 0 \text{ for } s > 0,\ f(s) \le a_1 + a_2 s^\delta, \text{ for some } a_1, a_2, \delta \ (a_1, a_2, \delta \text{ constants}). \tag{H_1}$$

Actually, in (Fraenkel and Berger, 1974) it is solved a slightly different problem : namely, instead of considering λ (the vortex-strength parameter) as a given constant, the Authors seek a solution Ψ of a given energy and find λ as a Lagrange multiplier. The procedure in (Fraenkel and Berger, 1974) is, roughly, the following :

i) In a first instance, the problem is substituted by a new one in a bounded domain : put $D = \{(r,z) : 0 < r < a,\ |z| < b\}$ and denoted by $-\chi(r)$ the stream function of the flow $(0,0,-W)$, i.e. $\chi(r) = \frac{1}{2}Wr^2 + k$, Π is replaced by D in (1a-b-c), while (1d-e) are substituted by the boundary condition $\Psi|_{\partial D} = -\chi$.

ii) The new free boundary problem, say $(1)_D$, is transformed into a semilinear Dirichlet problem. For this, it is sufficient to extend f to negative values setting $\overline{f}(s) = f(s)$ for $s > 0$ and $\overline{f}(s) = 0$ for $s \le 0$ and consider the equation $-L\Psi = \lambda r^2 \overline{f}(\Psi)$ which is equivalent, by the maximum principle to (1a-b). Now $A = \{(r,z) : \Psi > 0\}$. Writing $\Psi(r,z) = \psi(r,z) - \chi(r)$ and $g(r,s) = \overline{f}(s - \chi(r))$, it turns out that $(1)_D$ is equivalent to

$$\begin{aligned} -L\psi &= \lambda r^2 g(r,\psi) \text{ in } D \\ \psi &= 0 \qquad \text{on } \partial D \end{aligned} \tag{2}$$

iii) Problem (2) is solved by the calculus of variations in the large. For $u,v \in C_0^\infty(D)$, set $\langle u,v \rangle = \int_D r^{-2}(u_r v_r + u_z v_z) d\rho$, $d\rho = r\, dr\, dz$, and $\|u\|^2 = \langle u,u \rangle$; if H denotes the Hibert space obtained as closure of $C_0^\infty(D)$ with respect to the norm $\|.\|$, it results $H \hookrightarrow W_0^{1,2}(D)$ and $\langle u,v \rangle = -\int r^{-2} uLv\, d\rho$. Denote $G(u) = \int_0^u g(r,s)ds$. The weak solutions of (2) are obtained as critical points of $\int G(u)\, d\rho$ on the sphere $\|u\| = 1$, λ playing the rôle of a Lagrange multiplier. In particular $\int G(u) d\rho$ attains its maximum on the unit sphere at a point $\psi = \psi_D$.

iv) Since ψ maximizes $\int G(u) d\rho$ on $\|u\|^2 = 1$, it is shown that also ψ^*, the Steiner symmetrization of ψ with respect to the line $z = 0$ in D, maximizes $\int G(u)$ over $\|u\| = 1$, and thus it can be assumed $\psi = \psi^*$ (for definition and properties of the Steiner symmetrization, see (Hardy, Littlewood and Polya, 1952) or (Fraenkel and Berger, 1974)).

v) The property $\psi = \psi^*$ enables one to prove that the set $A = \{(r,z) : \psi > \chi\}$ is contained in an open subset $\Omega \subset \Pi$ which is independant of D.

vi) Lastly, taking a sequence $a_j \to +\infty$ and domains $D_j =]0,a_j[\times]-a_j,a_j[$, it is shown that ψ_{D_j} converges in $C^1(\overline{\Omega})$ to some $\overline{\psi}$; this $\overline{\psi}$ can be extended to a function ψ defined on all Π, in such a way that $\Psi \equiv \psi - \chi$ is the desired solution of (1).

(B) Solution of problem (1) with λ fixed.

Our purpose here is to study the free boundary problem (1) requiring that λ is fixed. Repeating the procedure in (Fraenkel and Berger, 1974), we deal with (2) : now the solutions of (2) are the (free) stationary points of $J(u) = \frac{1}{2}\|u\|^2 - \lambda\int_D G(u)d\rho$ on H. It is evident that only steps iii) and iv) before need a different proof : the remaining v-vi) can be repeated aferwards without any change. We will sketch below these arguments.

Let us assume that f satisfies (H_1) and

$$f(s) \text{ is convex for } s > 0 \qquad (H_2)$$

$$\exists\theta \in [0,\tfrac{1}{2}[\text{ such that } F(s) := \int_0^s f(\sigma)d\sigma \leq \theta s f(s) \quad \forall s \geq 0. \qquad (H_3)$$

Condition (H_3) implies f is superlinear at infinity.

Using a device as in (Coffman, 1969 and Hempel, 1971), we substitute the free critical point problem with an isoperimetric one.

We begin setting $\Gamma(u) = \Gamma_g(u) := \|u\|^2 - \lambda\int ug(u)d\rho$, $J(u) = J_g(u) := \frac{1}{2}\|u\|^2 - \lambda\int G(u)d\rho$, $M = M_g := \{u \in H : u \not\equiv 0 \text{ and } \Gamma(u) = 0\}$.

We remark that, since $\overline{f}(s) = 0 \quad \forall s < 0$, it follows that $u \in M$ only if the set $\{(r,z) : u(r,z) > 0\}$ has positive measure.

We first give some properties of M and J which easily follow from $(H_1\text{-}H_3)$ (see (Ambrosetti, 1972) for a proof, under slightly more restrictive conditions).

LEMMA 1. 1) Let $u \in H$ be such that meas $\{(r,z) : u(r,z) > 0\} > 0$. Then, set $\Gamma(tu) = t^2\gamma(tu)$, the function $t \to \gamma(tu)$ is decreasing for $t > 0$.

2) It $u \in M$, then $t \to J(tu)$ is increasing for $t \in [0,1]$.

3) If $u \in M$, then $\|u\|^2 \le \beta J(u)$ with $\beta = (\frac{1}{2} - \theta)^{-1}$.

We notice that from Lemma 1-1) and (H_3) it follows that $M \ne \emptyset$.

Now, in order to find critical points of $J|_M$ we need M to be a C^1 manifold, and thus we assume, in a first instance, that $\bar{f} \in C^2(\mathbb{R})$. This assumption will subsequently be eliminated by means of a limiting procedure. Indeed, with such regularity on $\bar{f}$, and in view of the remark preceding Lemma 1, M is a C^1 manifold. Moreover, the critical points of $J|_M$ are stationary points of J on H. Lastly, using also Lemma 1-3), it follows that $J|_M$ attains its minimum at a point ψ, with $J(\psi) > 0$ and $\psi > 0$ in D. For more details, see (Ambrosetti, 1972).

Next, we prove our main Lemma.

LEMMA 2. Let ψ be a minimizing point of $J|_M$. Then $\psi^* \in M$ and $J(\psi^*) = \min_M J$.

Proof. Since $J(\psi) \ge J(\psi^*)$ (see Hardy, Littlewood and Polya, 1952 and Fraenkel and Berger, 1974), it is sufficient to prove that $\psi^* \in M$. We have

$$\Gamma(\psi^*) = \|\psi^*\|^2 - \lambda \int \psi^* g(\psi^*) d\rho \le \|\psi\|^2 - \lambda \int \psi g(\psi) d\rho = \Gamma(\psi) = 0.$$

From Lemma 1-1) and recalling that $\psi^* \ge 0$ in D we infer : $\exists\, 0 < t^* \le 1$ such that $t^* \psi^* \in M$. From Lemma 1-2) it follows $J(t^* \psi^*) \le J(t^* \psi) \le J(\psi)$, with strict inequality whenever $t^* < 1$. Since ψ minimizes $J|_M$, then $t^* = 1$.

In order to eliminate the assumption $\bar{f} \in C^2(\mathbb{R})$, we take a sequence $g_n \ge g$, uniformly converging to g on compact sets, with uniform Lipschitz constant on bounded sets, and such that the above arguments can be applied.

We set $\Gamma_n \equiv \Gamma_{g_n}$, $M_n \equiv M_{g_n}$, $J_n \equiv J_{g_n}$ and indicate with ψ_n a point of minimum of $J_n|_{M_n}$.

<u>LEMMA 3</u>. $\sup_n \|\psi_n\| < +\infty$.

<u>Proof</u>. Let us take a function $h \leq g_n$ for every n and such that Lemmas 1 and 2 hold with $h \equiv g$. Fixed $v \in M_h$, $v > 0$ in D, since $\Gamma_n(v) \leq \Gamma_h(v) = 0$, by 1-1) it follows : $\exists t_n \leq 1$ such that $t_n v \in M_n$. Therefore, using now Lemma (1-2-3), and since $h \leq g_n$ implies $J_n(u) \leq J_h(u)$,
we infer

$$\|\psi_n\|^2 \leq \beta J_n(\psi_n) \leq \beta J_n(t_n v) \leq \beta J_h(t_n v) \leq \beta J_h(v) \equiv \text{const.}$$

We are now in position to state a first result concerning problem (2).

<u>THEOREM 4</u>. Let f satisfy $(H_1 - H_2 - H_3)$. Then (2) has a solution ψ_D, which minimizes $J|_M$ and is Steiner symmetric. Moreover, A_{ψ_D} is connected.

<u>Proof</u>. Lemma 3, by using regularity results as in (Fraenkel and Berger, 1974), allows to show that ψ_n C^1-converges to some $\psi = \psi_D$, which solves (2). Such a ψ is Steiner symmetric because ψ_n is.

In order to prove that $J(\psi) \leq J(u)$ $\forall u \in M$, we notice thate $g_n \geq g$ implies $\Gamma_n(u) \leq \Gamma(u)$, $J_n(u) \leq J(u)$ $\forall u \in H$. Now, let $u \in M$. Lemma 1-1) yields $t_n u \in M_n$ for some $t_n \leq 1$ and hence, using Lemma 1-2), $J(u) \geq J(t_n u)$ $\geq J_n(\psi_n)$. Since $J_n(\psi_n) \to J(\psi)$, the result follows.

Lastly, let us assume, by contradiction, that $A = A_1 \cup A_2$, A_i disjoint open sets. Set $\phi_i = \psi - \chi$ on A_i and $\phi_i = 0$ in $D \setminus A_i$, $i = 1,2$. Thus

$\phi_i \in H(D)$ and, taking in (2) the scalar product with ϕ_i, one easily obtains $\phi_i \in M$. But then, from $J(\psi) = J(\phi_1) + J(\phi_2)$ and $J(\phi_i) > 0$, we get a contradiction.

We wish now to perform, as in (Fraenkel and Berger, 1974) the steps v-vi) recalled before, to obtain a solution of (1). To do this, using the same notations as in vi), and denoting by ψ_{D_j} the solution of (2) with $D = D_j$ given by Theorem 4, we only need a uniform bound on $\|\psi_{D_j}\|$. This will be accomplished by the following.

LEMMA 5. $\sup \|\psi_{D_j}\| < +\infty$.

Proof. Set $v_j = \psi_{D_j}$ in D_1 and $v_j = 0$ in $D_j \setminus D_1$. Obviously $v_j \in H(D_j)$ and $v_j \in M_j$, where M_j is the manifold M_g in $H(D_j)$ (similar meaning for J_j). Since ψ_{D_j} minimizes J_j over M_j, and using Lemma 1-3), we obtain

$$\|\psi_{D_j}\|^2 \leq \beta J_j(\psi_{D_j}) \leq \beta J_j(v_j) = J_1(\psi_{D_1}) = \text{const.}$$

Using Theorem 4 and Lemma 5 we now see that (1) is solvable. In order to obtain a solution of physical interest, it remains to show that, if ψ is the C^1-limit of ψ_{D_j}, then $A_\psi \neq \emptyset$. Taking into account that $\psi(0,z) = 0 \ \forall z$ and the behaviour of ψ at infinity, an application of the maximum principle shows that it is enough to prove that $\psi \not\equiv 0$. Indeed, since $\psi_{D_j} \to \psi$ in $C^1(\Omega)$, if $\psi \equiv 0$, we would have $\psi_{D_j} < k$ $(r,z) \in \Omega$ and j large. Since $g(s,r) \equiv 0$ for $s < k$, it follows $\|\psi_{D_j}\|^2_{H(D_j)} = \int_{D_j} \psi_{D_j}(\psi_{D_j}) = 0$, a contradiction.

Hence we can state our main result.

THEOREM 6. Let f satisfy $(H_1\text{-}H_2\text{-}H_3)$. Then (1) has a solution with the property that the set $A = \{(r,z) : \psi(r,z) > 0\}$ is nonempty, bounded and symmetric with respect $z = 0$.

2. A FREE BOUNDARY PROBLEM ARISING IN PLASMA PHYSICS

The equilibrium of a plasma confined in a toroidal cavity, whose meridian section is the bounded domain Ω, can be described as follows (see Temam, 1975).

Given the bounded domain $\Omega \subset \mathbb{R}^N$, $I, \lambda > 0$ and $f : \mathbb{R}^+ \to \mathbb{R}^+$ with $f(0) = 0$, to find $k \in \mathbb{R}$, an open subset Ω_p of Ω and a C^1 function u such that

$$
\begin{aligned}
-\Delta u &= \lambda f(u) && \text{in } \Omega_p \\
\Delta u &= 0 && \text{in } \Omega_v \equiv \Omega \setminus \Omega_p \\
u &= 0 && \text{on } \partial\Omega_p \\
u &= -k && \text{on } \partial\Omega \\
-\int_{\partial\Omega} \frac{\partial u}{\partial n}\, d\sigma &= I
\end{aligned}
\qquad (3)
$$

The region Ω_p is the part of Ω filled by the plasma, Ω_v corresponds to the vacuum and n denotes the outer unit normal at $\partial\Omega$.

In the case $f(s) = s$ (the "model" case) problem (3) has been studied by several people (Puel, 1977 and Temam, 1977; among others); for general nonlinearities we refer to (Berestycki and Brezis, 1976).

Our result is

THEOREM 7. Let us assume :

$$f : \mathbb{R}^+ \to \mathbb{R}^+ \text{ is continuous and } f(0) = 0 \qquad (H')$$

$$s^{-p} f(s) \to 0 \text{ as } s \to +\infty \text{ with } p = \frac{N}{N-2} \text{ for } N > 2 \text{ and some } p \text{ for } N = 2 \qquad (H'')$$

Let $f_+ = \liminf_{s \to +\infty} f(s)$ ($f_+ = +\infty$ is allowed). Then

i) if $\int f_+ = +\infty$, (3) has solution for every $I > 0$;

ii) if $\int f_+ < +\infty$, $\exists b \geq \lambda \int f_+$ such that (3) has solution for all $0 < I < b$.

Moreover, if

$$\lim_{s \to +\infty} f(s) = f_+ < +\infty \text{ exists and } \exists \bar{s} > 0 \text{ such that } f(s) > f_+ \ \forall s \geq \bar{s} \qquad (H''')$$

then there exists $b > \lambda \int f_+$ such that (3) has two distinct solutions for all $\lambda \int f_+ < I < b$.

Theorem 7 improves the preceding works (Berestycki and Brezis, 1976 and to appear; Puel, 1977; Temam, 1977); the multiplicity result is new (Schaeffer has shown the existence of multiple solutions for $f(s) = s$ (Schaeffer, 1977).

The proof is based on a topological degree argument and global bifurcation techniques. Roughly, we proceed as follows :

1) We eliminate the unknown set Ω_p, extending, as in the first section, f to $\bar{f}$ defined by $\bar{f}(s) = 0 \ \forall s < 0$. Afterwards, setting $v = u+k$, (3) is transformed into

$$-\Delta v = \lambda \bar{f}(v-k) \quad \text{in } \Omega \qquad 4a)$$

$$v = 0 \quad \text{on } \partial\Omega \qquad 4b)$$

$$-\int_{\partial\Omega} \frac{\partial v}{\partial n} \, d\sigma = I \qquad 4c)$$

Ω_p will be the set $\{v > k\}$.

We denote by Σ the closure in $R \times W_0^{1,2}(\Omega)$ of the solutions (k,v), $v \not\equiv 0$ of (4a-b). Notice that, by the maximum principle, we have $v > 0$ in Ω.

2) Looking at k as a bifurcation parameter (which appears in a nonlinear manner in the equation), we show that there exists a connected component Σ^* of Σ such that (i) Σ^* is unbounded in $R \times W_0^{1,2}(\Omega)$, and (ii) $(0,0) \in \Sigma^*$ and $(k,0) \in \Sigma^*$ implies $k = 0$.

3) For $(k,v) \in \Sigma$, we set $J(k,v) = -\int_{\partial\Omega} \frac{\partial v}{\partial n} = \int_\Omega f(v-k)$ and $b := \sup_\Sigma J(k,v)$. Clearly, since Σ is connected, $(0,0) \in \Sigma^*$ with $J(0,0) = 0$, and $(k,v) \mapsto J(k,v)$ is continuous, then for every $0 < I < b$ the equation $J(k,v) = I$ has a solution, which is a solution of (3).

4) We estimate b. This is accomplished by the following Lemma.

LEMMA 8. Let $\Sigma' \subset \Sigma$ be such that $\sup_{\Sigma'} J(k,v) < +\infty$. Then for all $\bar{k} \in \mathbb{R}$ there exists $c > 0$ such that $\|v\|_{L^\infty} \le c$ for every $(k,v) \in \Sigma'$, $k \ge \bar{k}$.

An application of Fatou's Lemma, jointly with Lemma 8, implies $b > \lambda\int f_+$.

5) A bit more careful analysis of the behaviour of Σ^* and J permits us to show the multiplicity result, provided (H''') holds.

An important question is to know whether or not there is free boundary, namely when Ω_p is properly contained in Ω. This is equivalent to know the sign of k, the boundary value on $\partial\Omega$ of the solution of (3). An answer is given in the following result.

THEOREM 9. (i) Let us assume $f(s) \leq \alpha s \ \forall s > 0$. Then $\exists \underline{\lambda} > \lambda_1/\alpha$ such that $\forall \lambda \leq \underline{\lambda}$, all solutions of (3) satisfy $u|_{\partial\Omega} \geq 0$ (i.e. there is no free boundary)

(ii) If $f(s) \geq \beta s$, $\forall s > 0$, then $\exists \overline{\lambda} \leq \lambda_1/\beta$ such that $\forall \lambda > \overline{\lambda}$, all solutions of (3) satisfy $u|_{\partial\Omega} < 0$ (i.e. there is free boundary).

The above Theorem can be considered an improvement of results in (Puel, 1977), (Temam, 1977) for the "model case". In fact, if $\overline{f}(s) = s^+$, then $\underline{\lambda} = \overline{\lambda} = \lambda_1$. Besides such a case, in general one has $\underline{\lambda} < \lambda < \overline{\lambda}$ and it could be of some interest to remark that for $\underline{\lambda} < \lambda < \overline{\lambda}$ the existence of the free boundary depends on I (see Ambrosetti and Mancini, to appear).

Bibliography

1 A. Ambrosetti, Esistenza di infinite soluzioni per problemi non lineari in assenza di parametro. Atti Acc. Naz. dei Lincei, 52 (1972) 660-667.

2 A. Ambrosetti and G. Mancini, A free boundary problem and a related semilinear equation. To appear in Nonlinear Analysis T.M.A.

3 H. Berestycki and H.Brézis, Sur certains problèmes de frontière libre. Comptes Rendus Ac. Sc. Paris, 283 série A, (1976), 1091-1094.

4 H. Berestycki and H. Brézis, On e free boundary problem arising in plasma physics. To appear in Nonlinear Analysis T.M.A.

5 J.L. Bona, D.K. Bose and R.E.L. Turner, Finite amplitude steady waves in stratified fluids, to appear.

6 C.V. Coffman, A minimum maximum principle for a class of non-linear integral equations. J. Analyse Math. 22, (1969) 391-419.

7 L.E. Fraenkel and M.S. Berger, A global theory of steady vortex rings in an ideal fluid. Acta Math. 132, (1974), 14-51.

8 B. Gidas, L. Niranberg and W.M. Ni, Symmetry and related properties via the maximum principle. To appear.

9 G.H. Hardy, J.E. Littlewoood and G. Polya, Inequalities. Cambridge, 1952.

10 J.A. Hempel, Multiples solutions for a class of nonlinear boundary value problems. Ind. Univ. Math. J.20. (1971) 983-999.

11 J.P. Puel, Sur un problème de valeur propre non linéaire et de frontière libre. Comptes Rendus As. Sc. Paris, 284 Série A, (1977), 861-863.

12 D.G. Schaeffer, Non uniqueness in the equilibrium shape of a confined plasma. Comm. in Part. Diff. Equat., 2 (6), (1977), 587-600.

13 R. Temam, A nonlinear eigenvalue problem : The shape at equilibrium of a confined plasma. Arch. Rat. Mech. Anal. 60, (1975) 51-73.

14 R. Temam, Remarks on a free boundary value problem arising in plasma physics. Comm. in Part. Diff. Equations, 2 (1977), 563-585.

Antonio AMBROSETTI(*)

Istituto Matematico,
Università di Ferrara
44100 Ferrara
and
S.I.S.S.A., Strada Costiera 11,
34014 Trieste,

Italy

Giovanni MANCINI (*)

Istituto Matematico,
Università di Bologna
40100 Bologna

Italy

(*) work supported in part by CNR-GNAFA

JOHN M BALL

Les systèmes asymptotiquement dynamiques

1.

Commençons par un exemple simple. On considère l'équation différentielle ordinaire

$$\ddot{u} + \dot{u} + u^3 - u = f(t) \tag{1}$$

où f est continue et $f(t) \to 0$ quand $t \to \infty$. On ne suppose pas que

$$\int_{t_0}^{\infty} |f(t)|\, dt < \infty .$$

L'équation autonome correspondant à (1) est

$$\ddot{u} + \dot{u} + u^3 - u = 0. \tag{2}$$

Il est bien connu que toute solution $(u,\dot{u})$ de (2) tend quand $t \to \infty$ vers un des points stationnaires $(-1,0)$, $(0,0)$, $(1,0)$. On se pose la question : en va-t-il de même pour toute solution bornée de (1) ? La réponse à cette question est positive et ce résultat est une conséquence de la théorie décrite ci-dessous.

On note que dans notre exemple l'équation autonome (2) possède une fonction de Lyapunov

$$V(u,\dot{u}) = \frac{1}{2}\dot{u}^2 + \frac{1}{4}u^4 - \frac{1}{2}u^2, \tag{3}$$

mais que pour (1)

$$\dot{V}(u,\dot{u}) = f(t)\dot{u} - \dot{u}^2$$

n'a pas de signe défini.

2.

Soit (X,d) un espace métrique. Par un *processus* sur X (cf. essentiellement Dafermos [5]) on entend une famille d'opérateurs $U(t,s) : X \to X$, définis pour $t \in \mathbb{R}^+$, $s \in \mathbb{R}$, et satisfaisant

(i) $U(t,0)$ = identité,

(ii) $U(t+\tau,s) = U(t,s+\tau)U(\tau,s)$, $s \in \mathbb{R}$, $t,\tau \in \mathbb{R}^+$,

(iii) pour $s_o \in \mathbb{R}$, $t \in \mathbb{R}^+$ fixés, les opérateurs $U(t,s)$, avec paramètres $s \in [s_o,\infty)$, sont équicontinus (c'est-à-dire, si $\varepsilon > 0$, $x \in X$ sont donnés, il existe $\delta > 0$ tel que si $d(x,y) < \delta$ alors $d(U(t,s)x,U(t,s)y) < \varepsilon$ pour tout $s \in [s_o,\infty)$).

(Interprétation : $U(t,s)x$ est la valeur d'une solution au temps $s + t$ qui prend la valeur x au temps s.)

$U(.,.)$ est un *système asymptotiquement dynamique* s'il existe des opérateurs $T(t)$, $t \in \mathbb{R}^+$, tels que si $s_n \to \infty$, $t \in \mathbb{R}^+$, $x \in X$, alors

$$U(t,s_n)x \to T(t)x.$$

Il est clair que $T(.)$ est un *semi-groupe*, c'est-à-dire

(i) $T(0)$ = identité

(ii) $T(s+t) = T(s)T(t)$, $s,t \in \mathbb{R}^+$,

(iii) chaque opérateur $T(t) : X \to X$ est continu.

Soit $x \in X$, $s \in \mathbb{R}$. On définit *l'orbite positif*

$$\mathcal{O}^+(x,s) = \bigcup_{t\geq 0} U(t,s)x,$$

et *l'ensemble ω-limite*

$$\omega(x,s) = \{y \in X : \text{il existe une suite } t_n \to \infty \text{ telle que } U(t_n,s)x \to y\}.$$

Un ensemble $A \subset X$ est *invariant* si et seulement si $T(t)A = A$ pour tout $t \in \mathbb{R}^+$. Un ensemble *invariant* qui consiste en un seul point est un *point stationnaire*. Le résultat suivant est bien connu (cf. Dafermos [5]).

<u>LEMME 1</u> : Si $\mathcal{O}^+(x,s)$ est relativement compact, alors $\omega(x,s)$ est non vide, invariant et compact, et $d(U(t,s)x,\omega(x,s)) \to 0$ quand $t \to \infty$. Si de plus l'application $t \to U(t,s)x$ est continue pour t assez grand, alors $\omega(x,s)$ est connexe.

<u>DEMONSTRATION</u> : Nous donnons seulement la démonstration du fait que $\omega(x,s)$ est positif invariant, c'est-à-dire

$$T(t)\omega(x,s) \subset \omega(x,s), \quad t \in \mathbb{R}^+.$$

Soit $t_n \to \infty$, $U(t_n,s)x \to y \in \omega(x,s)$. Alors, par l'équicontinuité,

$$U(t_n+t,s)x = U(t,t_n+s)U(t_n,s)x \to T(t)y.$$

Donc $T(t)y \in \omega(x,s)$. □

Soit $V : X \to \mathbb{R}$ une fonction continue de *Lyapunov* pour $T(.)$; i.e.

$$V(T(x)) \leq V(x), \quad t \in \mathbb{R}^+, \quad x \in X.$$

Pour tout $\gamma \in \mathbb{R}$ soit

$$M_\gamma = \{x \in X : V(T(t)x) = \gamma \text{ pour tout } t \in \mathbb{R}^+\}$$

THEOREME 2 : Soit $x \in X$, $s \in \mathbb{R}$. Supposons que $\mathcal{O}^+(x,s)$ soit relativement compact.

Soit

$$\alpha = \underline{\lim}_{t\to\infty} V(U(t,s)x), \qquad \beta = \overline{\lim}_{t\to\infty} V(U(t,s)x).$$

Alors pour tout $\gamma \in [\alpha,\beta]$ l'ensemble $\omega(x,s) \cap M_\gamma$ est non vide.

COROLLAIRE 3 : Si de plus on suppose que l'application $t \to U(t,s)x$ est continue pour t assez grand et que $\bigcup_{\gamma\in[\alpha,\beta]} M_\gamma$ est fini, alors $\alpha = \beta$ et il existe un point stationnaire $\phi \in M_\alpha$ avec $U(t,s)x \to \phi$ lorsque $t \to \infty$.

DEMONSTRATION DU COROLLAIRE : Puisque $\bigcup_{\gamma\in[\alpha\ \beta]} M_\gamma$ est fini, $\alpha = \beta$. Mais puisque V est continue et $\omega(x,s)$ est invariant, $\omega(x,s) \subset M_\alpha$. Parce que $\omega(x,s)$ est connexe par le Lemme 1, on déduit que $\omega(x,s) = \{\phi\}$ où ϕ est un point stationnaire. □

EXEMPLE : X = le cercle du rayon unité = $\mathbb{R}$ (mod 2π).

On considère l'équation

$$\dot{\theta}(t) = -g(\theta) - f(t),$$

avec

$$g(\theta) = \begin{cases} \cos^2\theta\,, & -\frac{\pi}{2} \le \theta \le \frac{\pi}{2}, \\ 0, & \frac{\pi}{2} \le \theta \le \frac{3\pi}{2}\,, \end{cases}$$

et avec $f(t) \ge 0$, f continue, $f(t) \to 0$ lorsque $t \to \infty$, mais $\int_0^\infty f(t)dt = \infty$.

Les points stationnaires de l'équation autonome

$$\dot{\theta}(t) = -g(\theta)$$

sont $\{\theta: \frac{\pi}{2} \leq \theta \leq \frac{3\pi}{2}\}$. La fonction $V(\theta) = \sin \theta$ est une fonction de Lyapunov, et chaque M_γ, $\gamma \in [-1,1]$, contient un seul point. Mais en intégrant (4),

$$\theta(t) - \theta(0) \leq - \int_0^t f(s)ds \to -\infty \quad \text{quand} \quad t \to \infty .$$

Donc $\omega(\theta_o, 0) = X$ ne contient pas que des points stationnaires.

Mais la conclusion du Théorème 2 est valable (voir Figure 1).

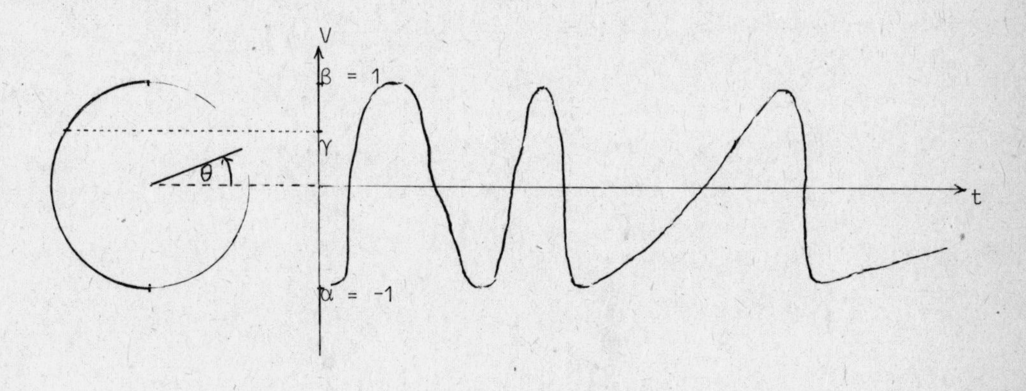

Figure 1

On note dans cet exemple que chaque fois que $V(t)$ croît elle croît plus lentement. Ce fait est vrai en général et est utilisé dans la démonstration du théorème.

DEMONSTRATION DU THEOREME 2 : Soit $V(t) = V(U(t,s)x)$. On a que pour tout $\tau \in \mathbb{R}^+$

$$\underline{\lim}_{t\to\infty} V(t) - V(t+\tau) \geq 0 .$$

Sinon il existerait $t_n \to \infty$ avec

$$V(t_n) - V(t_n+\tau) \leq \varepsilon < 0 \quad \text{pour tout} \quad n,$$

et sans restreindre la généralité,

$$U(t_n,s) \to y, \; U(t_n+\tau,s)y \to T(\tau)y.$$

Alors

$$V(y) - V(T(\tau)y) \leq \varepsilon < 0,$$

d'où la contradiction.

On a besoin du lemme suivant, qui est démontré dans [2] grâce au théorème de Baire.

LEMME 4 : Soit $f : \mathbb{R}^+ \to \mathbb{R}$ continue et pour tout $\tau \in \mathbb{R}^+$

$$\underline{\lim}_{t\to\infty} f(t) - f(t+\tau) \geq 0.$$

Soit $\alpha = \underline{\lim}_{t\to\infty} f(t)$, $\beta = \overline{\lim}_{t\to\infty} f(t)$ avec $-\infty < \alpha \leq \beta < \infty$. Alors pour tout $\gamma \in [\alpha,\beta]$ il existe une suite $t_n \to \infty$ telle que

$$f(t_n+t) \to \gamma$$

uniformément pour t dans tout sous-ensemble compact de $\mathbb{R}^+$.

En appliquant le Lemme 4 à V, on obtient pour tout $\gamma \in [\alpha,\beta]$ une suite $t_n \to \infty$ telle que $U(t_n,s)x \to y \in \omega(s,x)$ et $V(U(t_n+t,s)x) \to \gamma$ pour tout $t \in \mathbb{R}^+$. Donc $V(T(t)y) = \gamma$ pour tout $t \in \mathbb{R}^+$ et $y \in \omega(s,x) \cap M_\gamma$.
□

EXEMPLES : I. Considérons à nouveau l'équation (1). Prenons $X = \mathbb{R}^2$, $z = \begin{bmatrix} u \\ \dot{u} \end{bmatrix}$. Alors (1) devient

$$\frac{dz}{dt} = \begin{bmatrix} \dot{u} \\ -\dot{u}-u^3+u \end{bmatrix} + \begin{bmatrix} 0 \\ f(t) \end{bmatrix}. \qquad (7)$$

Soit

$$U(t,s) \begin{bmatrix} u_o \\ v_o \end{bmatrix} = z(s+t),$$

où $z : [s,\infty) \to X$ est la solution de (7) avec $z(s) = \begin{bmatrix} u_o \\ v_o \end{bmatrix}$. On peut prouver aisément que les opérateurs $U(t,s)$ forment un système asymptotiquement dynamique, l'existence globale des solutions résultant de l'estimation

$$\dot{V} + u^2 = f(t)\dot{u} \leq \frac{1}{2} f(t)^2 + \frac{1}{2} \dot{u}^2.$$

Le semi-groupe correspondant est donné par

$$T(t) \begin{bmatrix} u_o \\ v_o \end{bmatrix} = z(t),$$

où $z(t)$ est la solution de

$$\frac{dz}{dt} = \begin{bmatrix} \dot{u} \\ -\dot{u}-u^3+u \end{bmatrix} ; \quad z(0) = \begin{bmatrix} u_o \\ u_o \end{bmatrix}.$$

Du Théorème 2 on déduit que toute solution, bornée dans X, de (7) tend vers un des points stationnaires $u = 0$, $u = \pm 1$ quand $t \to \infty$.

Remarque : Suivant Artstein [1] on peut affaiblir les conditions sur f.

Soit $f \in L^1_{loc}(\mathbb{R})$ avec $\int_{a+t}^{b+t} f(\tau)d\tau \to 0$ quand $t \to \infty$ pour tout $a, b \in \mathbb{R}$.

Alors toute solution bornée tend vers un point stationnaire.

II - Considérons le problème

$$\begin{cases} u_t = u_{xx} + f(x,t,u), & 0 < x < 1, t > s, \\ u_x(0,t) = u_x(1,t) = 0, & t > s, \\ u(x,s) = \psi(x), & \end{cases} \tag{8}$$

où $\psi \in C([0,1])$. Le problème autonome correspondant est

$$\begin{cases} \overline{u}_t = \overline{u}_{xx} + \overline{f}(x,\overline{u}), \\ \overline{u}_x(0,t) = \overline{u}_x(1,t) = 0, \\ \overline{u}(x,0) = \overline{\psi}(x). \end{cases} \tag{9}$$

Supposons que

(i) $f, \overline{f}$ soient régulières

(ii) il existe des constantes $a(s)$, $M(\rho,s)$ telles que

$$\sup_{x \in [0,1]} vf(x,t,v) \leq 0 \quad \text{pour} \quad |v| \geq a(s), \quad t \geq s,$$

$$\|f(.,t,v)\|_{C^2([0,1])} \leq M(\rho,s) \quad \text{pour} \quad |v| \leq \rho, t \geq s.$$

(iii) pour tout $\rho > 0$

$$\lim_{t\to\infty} \int_t^{t+1} \sup_{|v| \leq \rho} \|f(.,\tau,v) - \overline{f}(.,v)\|_{C([0,1])} d\tau = 0.$$

Pour $u \in C^1([0,1])$ soit

$$V(u) = \int_0^1 \left[\frac{1}{2} u_x^2 - \overline{F}(x,u) \right] dx,$$

où

$$\overline{F}(x,u) \overset{\text{def}}{=} \int_0^u \overline{f}(x,s)ds.$$

Soit

$$U(t,s)\psi = u(.,t),$$

où u est la solution de (8) et $\psi \in C^1([0,1])$. Soit aussi

$$T(t)\overline{\psi} = \overline{u}(.,t),$$

où $\overline{u}$ est la solution de (9) et $\overline{\psi} \in C^1([0,1])$. Alors on peut démontrer que les opérateurs $U(t,s)$ forment un système asymptotiquement dynamique dans $X = C^1([0,1])$ avec le semi-groupe correspondant $T(t)$. De plus, si $\psi \in C([0,1])$ alors $U(t,s)\psi \in C^1([0,1])$ pour tout $t > 0$. Donc pour étudier le comportement asymptotique des solutions de (8) on peut supposer que $\psi \in C^1([0,1])$. On peut aussi prouver que toute solution de (8) est bornée dans $C([0,1])$ (par le principe du maximum) et donc que $\mathcal{O}^+(\psi,s)$ est relativement compact dans $C^1([0,1])$. En outre V est une fonction de Lyapunov continue, et les ensembles M_γ contiennent seulement des points stationnaires. Donc en appliquant le Théorème 2 et le Corollaire 3 on obtient

<u>THEOREME 5</u> : Soit u la solution de (8) et

$$\alpha = \underline{\lim}_{t\to\infty} V(u(.,t)), \quad \beta = \overline{\lim}_{t\to\infty} V(u(.,t))$$

Alors $\infty < \alpha \le \beta < \infty$ et pour tout $\lambda \in [\alpha,\beta]$ il existe un point stationnaire $w(.) \in \omega(\psi,s)$ (l'ensemble ω-limite dans $C^1([0,1])$) avec $V(w(.)) = \gamma$. Si de plus les points stationnaires du (9) sont isolés dans $C([0,1])$, alors $\alpha = \beta$ et quand $t \to \infty$ $u(.,t)$ tend dans $C^1([0,1])$ vers un seul point stationnaire $w(.)$ avec $V(w(.)) = \alpha$.

Remarque : Si les points stationnaires ne sont pas isolés, on ne peut pas conclure que $u(.,t)$ tend quand $t \to \infty$ vers un seul point stationnaire. Par exemple, soit $\bar{f}$ telle que $\bar{f}(r) = 0$ si $r \in [-1,1]$, $|\bar{f}(r)| \geq 1$ si $|r| \geq 2$, $f(x,t,r) = \bar{f}(r) + g(t)$, où $g(t) \to 0$ quand $t \to \infty$ mais $\overline{\lim}_{t\to\infty} \int_0^t g(s)ds = 1$, $\underline{\lim}_{t\to\infty} \int_0^t g(s)ds = -1$, $\sup_{t\geq 0} \left| \int_0^t g(s)ds \right| = 1$. Alors la solution de (8) avec $\psi \equiv 0$, $s = 0$, est

$$u(x,t) = \int_0^t g(s)ds,$$

et donc l'ensemble ω-limite est donné par

$$\omega(0,0) = \{w(.) \equiv c : c \in [-1,1]\} .$$

Cependant pour l'équation autonome (9), sous nos hypothèses, toute solution tend vers un seul point stationnaire, même si ces points ne sont pas isolés. Ceci est dû au fait que $T(.)$ est un semi-groupe de contractions dans $L^2(0,1)$ (voir Brezis [4]).

4. DISCUSSION : Pour une bibliographie récente concernant les applications des fonctions de Lyapunov aux équations différentielles voir Dafermos [6] .

Une version du Corollaire 3 a été donnée pour la première fois dans Ball et Peletier [3]. Dans cet article, une hypothèse d'équicontinuité des opérateurs $U(t,s)$ plus forte que dans le présent article a été faite ce qui évite l'usage du théorème de Baire dans le Lemme 4. On trouve aussi dans [3] une discussion de l'exemple II dans le cas où les points stationnaires de (9) sont isolés.

Dans le Théorème 2 il n'est pas essentiel d'avoir l'unicité des solutions. En outre il suffit de savoir seulement que la solution dont on désire étudier le comportement asymptotique existe pour tout $t \geq s$. De plus, si on examine la démonstration du Théorème 2 on voit qu'il n'est pas essentiel d'avoir V continue. Si V n'est pas continue, alors l'ensemble

$$\omega_V(x,s) = \{y \in X : \text{il existe une suite } t_n \to \infty \text{ telle que } U(t_n,s)x \to y \text{ et } V(U(t_n,s)x) \to V(y)\}$$

peut être différent de $\omega(x,s)$. Si (6) est vraie, et si les conditions

$$\begin{cases} U(t_n,s)x \to y, \\ V(U(t_n,s)x) - V(U(t_n+t,s)x) \to 0 \quad \text{uniformément pour } t \text{ appartenant aux compacts de } \mathbb{R}^+, \end{cases}$$

impliquent que

$$V(U(t_n,s)x) \to V(y),$$

alors la démonstration établit que

$$\omega_V(x,s) \cap M_\gamma \neq \phi \quad \text{pour tout } \gamma \in [\alpha,\beta].$$

Cette observation est importante pour les problèmes hyperboliques, où X peut être un sous-ensemble d'un espace d'Hilbert, muni de la topologie faible. Si, par exemple,

$$V(x) = \frac{1}{2}\,|x|_X^2 + g(x),$$

avec $g : X \to \mathbb{R}$ compacte, alors $\omega_V(x,s)$ est l'ensemble ω-limite fort.

En utilisant ces idées on peut obtenir le même type de résultat que dans le Théorème 5 pour les équations de type

$$\begin{cases} u_{tt} - \Delta u + u_t + f(u,t) = 0, & x \in \Omega, \\ u|_{\partial\Omega} = 0, \end{cases}$$

où $\Omega \subset \mathbb{R}^n$ est un ouvert borné. Voir [2] pour tous ces développements de la théorie.

Bibliographie :

1 Z. Artstein, The limiting equations of non-autonomous ordinary differential equations, J. Diff. Equations 25(1977) 184-202.

2 J.M. Ball, On the asymptotic behaviour of generalised processes, with applications to nonlinear evolution equations, J. Diff. Equations 27(1978) 224-265.

3 J.M. Ball & L.A. Peletier, Global attraction for the one-dimensional heat equation with nonlinear time-dependent boundary conditions, Arch. Rat. Mech. Anal. 65(1977) 193-201.

4 H. Brezis, Monotonicity methods, dans "Contributions to Non-linear Functional Analysis", édité par Zarantonello, Academic Press, New York 1971.

5 C.M. Dafermos, An invariance principle for compact processes. J. Diff. Equations, 9(1971) 239-252.

6 C.M. Dafermos, Asymptotic behaviour of solutions of evolution equations, dans "Non linear Evolution Equations" édité par M.G. Grandall, Academic Press, New York 1978.

J.M. BALL

Department of Mathematics
Heriot-Watt University

Edinburgh EH14 4AS

ECOSSE

JOHN M BALL

Remarques sur l'existence et la régularité des solutions d'elastostatique non linéaire

1.

Examinons tout d'abord le problème en dimension un ; soit

$$\frac{d}{dx} W_p(X,x'(X)) = \Phi_x(X,x(X)), \ 0 < X < 1, \qquad (1)$$

avec les conditions au bord

$$x(0) = 0, \ x(1) = \lambda > 0. \qquad (2)$$

Ici x(X) est la position déformée d'un point d'un corps élastique unidimensionel qui était à $X \in [0,1]$ dans une configuration de référence (cf. Fig.1).

configuration de référence configuration déformée

Figure 1

La contrainte W_p est la dérivée par rapport à p de la fonction d'énergie interne W(X,p), et la force externe Φ_x est la dérivée par rapport à x de la fonction Φ(X,x) d'énergie potentielle. L'équation (1) est l'équation d'Euler pour la fonctionelle

$$I(x) = \int_0^1 [W(X,x'(X)) + \Phi(X,x(X))]dX.$$

Si Φ est zéro et W = W(p) homogène, une condition nécessaire pour que toutes les solutions de (1) soient C^1 est que W' soit strictement monotone, parce que si $W'(p) = W'(q)$ avec $p \neq q$, alors

$$x(X) = pX, \quad 0 \leq X \leq \frac{1}{2},$$

$$= qX + \frac{1}{2}(p-q), \quad \frac{1}{2} \leq X \leq 1,$$

est une solution. Si W est bornée inférieurement, cela implique que W est strictement convexe.

Nous faisons les hypothèses suivantes sur W et Φ :

(H_1) $W : [0,1] \times (0,\infty) \to \mathbb{R}$ est une fonction de Carathéodory, c'est-à-dire

$W(.,p)$ est mesurable pour tout $p > 0$,

$W(X,.)$ est continue pour presque tout $X \in (0,1)$.

(H_2) $W(X,p) \to \infty$ quand $p \to 0+$ pour presque tout $X \in (0,1)$.

(On définit $W(X,p) = \infty$ si $p \leq 0$.)

(H_3) $W(X,p) \geq \psi(p) + a(X)$ pour tout $p > 0$ et presque tout $X \in (0,1)$, où $a(.) \in L^1(0,1)$ et $\psi : \mathbb{R}^+ \to \mathbb{R}^+$ est une fonction convexe telle que $\psi(0) = 0$, $\frac{\psi(s)}{s} \to \infty$ quand $s \to \infty$.

(H_4) $\int_0^1 W(X,\lambda)dX < \infty$ pour tout $\lambda > 0$.

(H_5) $\Phi : [0,1] \times \mathbb{R}^+ \to \mathbb{R}$ est une fonction de Carathéodory, et pour tout $\lambda > 0$ il existe une constante $C(\lambda)$ telle que

$$|\Phi(X,r)| \leq C(\lambda) \text{ pour tout } r \in [0,\lambda] \text{ et presque tout } X \in (0,1).$$

Théorème 1 : Supposons de plus que $W(X,.)$ est convexe pour presque tout $X \in (0,1)$. Alors il existe une fonction x qui minimise I dans

$$S = \{x \in W^{1,1}(0,1) : I(x) < \infty, x(0) = 0, x(1) = \lambda\}.$$

Démonstration : Puisque $\lambda X \in S$, S est non vide. Chaque fonction $x \in S$ est continue et satisfait $x' > 0$ p.p., et donc $0 < x(X) < \lambda$ si $X \in (0,1)$. Donc I est bornée inférieurement dans S. A cause de (H3) et du critère de la Vallée Poussin il existe une suite minimisante pour I dans S convergeant faiblement dans $W^{1,1}(0,1)$. Puisque I est faiblement semicontinue inférieurement on obtient le résultat (cf. Ekeland et Témam [7 Thm 2.1, p.226]). □

Remarque : Dans le cas où W est homogène, nos hypothèses sur W sont équivalentes à $W : (0,\infty) \to \mathbb{R}$ convexe, $W(p) \to \infty$ quand $p \to 0+$, et $W'(p) \to \infty$ quand $p \to \infty$.

La démonstration du fait que la fonction minimisante x satisfait l'équation d'Euler (1) n'est pas immédiate, $x'(X)$ pouvant être zéro sur un ensemble non vide. Cette difficulté a été traitée par Antman [1] et par Antman & Brezis [2] sous des hypothèses différentes. Signalons que dans la première partie du théorème suivant on ne suppose ni que $W(X,.)$ est convexe, ni que (H3) est satisfaite.

Théorème 2 : Supposons que (H1), (H2), (H5) soient satisfaites et

(H6) $\begin{cases} W(X,.) : (0,\infty) \to \mathbb{R},\ \Phi(X,.) : \mathbb{R}^+ \to \mathbb{R} \text{ sont } C^1 \text{ pour presque tout} \\ X \in (0,1)\ ;\ W_p,\ \Phi_x \text{ sont des fonctions de Carathéodory, et pour tout} \\ \lambda > 0 \text{ il existe une constante } D(\lambda) \text{ telle que} \\ \\ |\Phi_p(X,r)| \le D(\lambda) \text{ pour tout } r \in [0,\lambda] \text{ et presque tout } X \in (0,1). \end{cases}$

Soit x qui minimise I dans S. Alors (1) est satisfaite p.p. dans $(0,1)$.

Si de plus W_p est continue, si $W(X,.)$ est strictement convexe pour tout $X \in [0,1]$, et si

$$W(X,p) \geq \theta(p), \; X \in [0,1], \; p > 0,$$

où θ est convexe, $\theta(p) \to 0$ quand $p \to 0+$, $\frac{\theta(p)}{p} \to \infty$ quand $p \to \infty$, alors $x \in C^1([0,1])$ et

$$\min_{X \in [0,1]} x'(X) > 0.$$

<u>Démonstration</u> : Soit $\Omega_n = \{X \in [0,1] : \sup_{|r-x'(X)|<\frac{1}{n}} |W_p(X,r)| \leq n\}$ et soit χ_n la fonction caractéristique de Ω_n. Puisque W_p est de Carathéodory, Ω_n est mesurable. Il est clair que $\Omega_n \subset \Omega_{n+1}$ et que $x'(X) = 0$ pour presque tout $X \in [0,1] \setminus \bigcup_{n=0}^{\infty} \Omega_n$. Parce que $I(x) < \infty$ on déduit de (H2) que $\text{mes}([0,1] \setminus \bigcup_{n=1}^{\infty} \Omega_n) = 0$. Prenons $v \in L^\infty(0,1)$ avec $\int_{\Omega_n} v(X)dX = 0$. Pour $|t|$ suffisamment petit on définit $y \in W^{1,1}(0,1)$ par

$$y'(X) = x'(X) + t\chi_n(X)v(X), \; y(0) = 0.$$

Puisque

$$I(y) = I(x) + \int_{\Omega_n} [W(X,x'(X) + tv(X)) - W(X,x'(X))]dX + \int_0^1 [\Phi(X,y(X)) - \Phi(X,x(X))]dX,$$

il est clair que $y \in S$. En divisant par t et en faisant $t \to 0$, on obtient

$$\int_{\Omega_n} W_p(X,x'(X))v(X)dX + \int_0^1 \phi_x(X,x(X))\left[\int_0^X \chi_n(Y)v(Y)dY\right]dX = 0,$$

où on a utilisé le théorème de la convergence bornée. Une intégration par parties donne

$$\int_{\Omega_n} \left[W_p(X,x'(X)) - \int_0^X \Phi_x(Y,x(Y))dY \right] v(X)dX = 0.$$

Mais v est arbitraire, et donc

$$W_p(X,x'(X)) - \int_0^X \Phi_x(Y,x(Y))\, dY = C, \qquad (3)$$

p.p. dans Ω_n. Evidemment C est indépendant de n, et donc (3) est satisfaite p.p. dans (0,1).

Pour démontrer la deuxième partie du théorème il suffit de prouver que $X_n \to X$ dans [0,1], $W_p(X_n,p_n) = a_n \to a$ impliquent $p_n \to p$, où p est la solution unique de $W_p(X,p) = a$. C'est clair si $0 < \varepsilon \le p_n \le k < \infty$ pour tout n. Sinon on peut supposer que $p_n \to 0$ ou $p_n \to \infty$. Si $p_n \to 0$, alors pour n suffisamment grand la convexité de $W(X_n,.)$ donne

$$W_p(X_n,p_n) \le \frac{W(X_n,p_n) - W(X_n,1)}{p_n - 1},$$

et donc $a_n \to -\infty$. On obtient une contradiction semblable si $p_n \to \infty$. □

<u>Remarque</u> : Sans une condition de croissance sur W on ne peut pas conclure que $x \in C^1([0,1])$, même si W(X,.) est strictement convexe. Par exemple la solution de

$$\min_{x(0)=0,x(1)=1} \int_0^1 \left[\frac{1}{x'^2} - 48\sqrt{2}\, x\left|X - \frac{1}{2}\right|^{\frac{1}{2}} \operatorname{sgn}(X-\tfrac{1}{2})\right] dX$$

est

$$x(X) = \frac{1}{2} + \frac{1}{\sqrt{2}} \left|X - \frac{1}{2}\right|^{\frac{1}{2}} \operatorname{sgn}(X-\frac{1}{2}).$$

Considérons maintenant la question de l'existence d'une fonction minimisante quand $W(X, \cdot)$ n'est pas convexe. En général on n'a pas l'existence : par exemple

$$\inf_{x(0)=0, x(1)=\frac{3}{2}} \int_0^1 [\frac{(x'-1)^2(x'-2)^2}{x'} + (x-\frac{3}{2}X)^2]dX = 0,$$

mais $x(X) = \frac{3}{2}X$ n'est pas minimisante. Remarquons que l'intégrale dans cet exemple est égale à

$$-\frac{3}{2} + \int_0^1 (W(X,x') + x^2)dX,$$

où

$$W(X,x') = \frac{(x'-1)^2(x'-2)^2}{x'} + \frac{3}{2}X^2x'.$$

(Pour des exemples reliés voir Young [11]). Dans un article récent Aubert & Tahraoui [3] ont donné des conditions sous lesquelles on a l'existence. En utilisant leurs méthodes on peut démontrer le résultat suivant.

<u>Proposition 3</u> : Si $\Phi \equiv 0$ alors le Théorème 1 est valable sans l'hypothèse que $W(X,.)$ est convexe.

On peut déduire de cette proposition un théorème d'existence pour le problème homogène :

$$\min_{x \in S} I(x),$$

où

$$I(x) = \int_0^1 [W(x'(X)) + \Phi(x(X))]dX.$$

<u>Théorème 4</u> : Soient $W : (0,\infty) \to \mathbb{R}$ continue, $W(p) \to \infty$ quand $p \to 0+$ ($W(p) = \infty$ si $p \leq 0$), $\frac{W(p)}{p} \to \infty$ quand $p \to \infty$, $\Phi : \mathbb{R}^+ \to \mathbb{R}$ continue. Alors il existe une fonction x qui minimise I dans S. Si de plus W,Φ sont C^1 alors x satisfait

$$\frac{d}{dX}W'(x'(X)) = \Phi'(x(X)) \quad \text{p.p. dans } [0,1].$$

<u>Démonstration</u> : On définit la fonction

$$\hat{W}(p) = pW(\frac{1}{p}).$$

Il est clair que

$$\hat{W}(p) \to \infty \text{ quand } p \to 0+, \quad \frac{\hat{W}(p)}{p} \to \infty \text{ quand } p \to \infty.$$

On définit $\hat{W}(p) = \infty$ si $p \leq 0$. Si $x \in S$ alors $x'(X) > 0$ p.p. et x possède un inverse $X(.)$ continu. Donc notre problème est équivalent à

$$\min_{x\in\hat{S}} \hat{I}(X),$$

où

$$\hat{I}(X) = \int_0^\lambda [\hat{W}(X'(x)) + \Phi(x)X'(x)]dx,$$

$$\hat{S} = \{X \in W^{1,1}(0,\lambda) : \hat{I}(X) < \infty, X(0) = 0, X(\lambda) = 1\}.$$

On résoud ce problème en utilisant la Proposition 3. L'équation d'Euler est satisfaite grâce au Théorème 2.

<u>Remarques</u> : (i) Voir [3] pour un autre résultat applicable au cas homogène.

(ii) On trouve dans [5] des observations concernant la transformation (4).

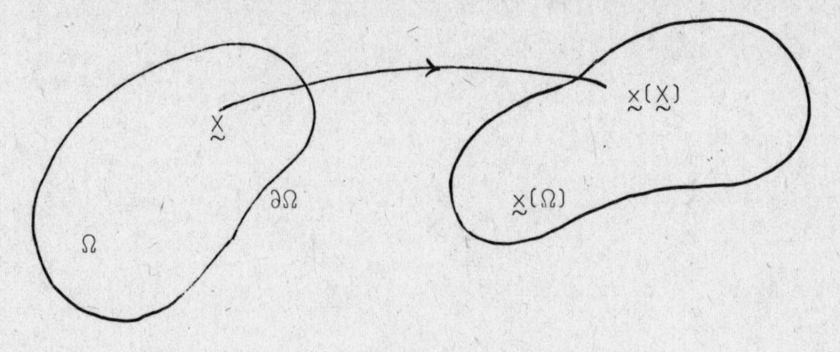

Figure 2

Dans le cas tri-dimensionel la déformation est donnée par une fonction $\underset{\sim}{x} : \Omega \to \mathbb{R}^3$, où $\Omega \subset \mathbb{R}^3$ est un ouvert borné de bord $\partial\Omega$ (cf. Figure 2).

Par raison de simplicité nous supposons qu'il n'y a aucune force externe et que le corps élastique est homogène. Alors les équations d'équilibre sont les équations d'Euler pour la fonctionelle

$$I(\underset{\sim}{x}) = \int_\Omega W(\nabla\underset{\sim}{x}(\underset{\sim}{X}))d\underset{\sim}{X},$$

c'est-à-dire

$$\frac{\partial}{\partial X^\alpha}\left(\frac{\partial W}{\partial x^i{}_{,\alpha}}\right) = 0, \; \underset{\sim}{X} \in \Omega, \; i = 1,2,3, \tag{5}$$

où W est la fonction d'énergie interne. Dans un problème aux limites de déplacement on doit trouver $\underset{\sim}{x}(.)$ satisfaisant (5) et

$$\underset{\sim}{x}(\underset{\sim}{X}) = \underset{\sim}{x}_0(\underset{\sim}{X}), \; \underset{\sim}{X} \in \partial\Omega, \tag{6}$$

où $\underset{\sim}{x}_0$ est une fonction donnée.

Suivant Hadamard [9] considérons un plan $\underset{\sim}{X}.\underset{\sim}{N} = k$ de $\mathbb{R}^3$. Il existe une fonction continue $\underset{\sim}{x}(\underset{\sim}{X})$ satisfaisant

$$\nabla\underset{\sim}{x}(\underset{\sim}{X}) = G \text{ quand } \underset{\sim}{X}.\underset{\sim}{N} < k,$$

$$\nabla\underset{\sim}{x}(\underset{\sim}{X}) = F \text{ quand } \underset{\sim}{X}.\underset{\sim}{N} > k,$$

où F,G sont des matrices constantes 3×3, si et seulement si il existe $\lambda \in \mathbb{R}^3$ tel que

$$F - G = \underset{\sim}{\lambda} \otimes \underset{\sim}{N}.$$

(Par définition, $(\underset{\sim}{\lambda} \otimes \underset{\sim}{N})^i = \lambda^i N_\alpha$). Une telle fonction $\underset{\sim}{x}$ est une solution faible de (5) si et seulement si le saut de la contrainte à travers le plan est zéro :

$$\left[\frac{\partial W}{\partial F^i_\alpha} N_\alpha\right] = \left[\frac{\partial W}{\partial F^i_\alpha}(G + \underset{\sim}{\lambda} \otimes \underset{\sim}{N}) - \frac{\partial W}{\partial F^i_\alpha}(G)\right] N_\alpha = 0. \tag{7}$$

Si W satisfait <u>la condition de l'ellipticité forte</u>

$$\frac{\partial^2 W(F)}{\partial F^i_\alpha \partial F^j_\beta} \lambda^i \lambda^j N_\alpha N_\beta > 0 \text{ si } \underset{\sim}{\lambda} \otimes \underset{\sim}{N} \neq 0 \tag{8}$$

alors

$$\left[\frac{\partial W}{\partial F^i_\alpha} N_\alpha\right] = \int_0^1 \frac{\partial^2 W}{\partial F^i_\alpha \partial F^j_\beta}(G + t\underset{\sim}{\lambda} \otimes \underset{\sim}{N})\lambda^i \lambda^j N_\alpha N_\beta dt = 0$$

implique que $\underset{\sim}{\lambda} \otimes \underset{\sim}{N} = 0$, c'est-à-dire F = G. Donc la condition (8) empêche les solutions $\underset{\sim}{x}$ de (5) pour lesquelles $\underset{\sim}{x}$ saute à travers un plan (cf. Knowles & Sternberg [10]). Pour une réciproque voir [12]. Pour des résultats d'existence de solutions de (5) sous des conditions impliquant (8) voir [4].

Question : Est-ce-que l'ellipticité forte implique que toute solution faible de (5) est C^1 ?

La réponse à cette question est non, et nous donnons un exemple ci-dessous. Pour le motiver rappelons un argument de [5,6]. Considérons un cube (du matériau de côté $\frac{1}{\lambda}$. Le cube est soumis à la déformation

$$\underset{\sim}{x}(\underset{\sim}{X}) = F(\lambda)\underset{\sim}{X},$$

où $F(\lambda)$ est une matrice 3×3 avec $|F(\lambda)| = \lambda$. Le diamètre du cube déformé est de l'ordre de l'unité. L'énergie de la déformation est égale à

$$\frac{W(F(\lambda))}{\lambda^3}.$$

Donc on peut obtenir à partir d'un cube infinitésimal une ligne de longueur 1 avec énergie finie si et seulement si

$$\frac{W(F)}{|F|^3} \not\to \infty \text{ quand } |F| \to \infty. \qquad (9)$$

Une fonction d'énergie interne satisfaisant (8) et (9) est

$$W(F) = tr(FF^T) + h(\det F), \qquad (10)$$

où $h : (0,\infty) \to \mathbb{R}$ est convexe, $h(1) = 0$, $h(\delta) \to \infty$ quand $\delta \to 0+$, $\frac{h(\delta)}{\delta} \to \infty$ quand $\delta \to \infty$.

Soient $\Omega = \{\underset{\sim}{X} : |\underset{\sim}{X}| < 1\}$, $R = |\underset{\sim}{X}|$, $r = |\underset{\sim}{x}|$. Considérons les déformations radiales ayant la forme

$$\underset{\sim}{x}(\underset{\sim}{X}) = \frac{r(R)}{R}\underset{\sim}{X}. \qquad (11)$$

Alors

$$I = I(r) = 4\pi \int_0^1 \left[R^2 r'^2 + 2r^2 + R^2 h\left(\frac{r^2 r'}{R^2}\right)\right] dR,$$

et (5) réduit à une équation différentielle ordinaire pour $r(R)$. On considère la condition au bord

$$r(1) = \lambda > 0. \tag{12}$$

On a toujours la solution triviale $r = \lambda R$, pour laquelle $\nabla \underset{\sim}{x} = \lambda I$. On peut démontrer que λR est la seule solution radiale si $\lambda \leq 1$, mais que si $\lambda > 1$ est suffisamment grand alors λR ne minimise pas $I(r)$ et la fonction $\bar{r}(R)$ minimisante satisfait $\bar{r}(0) > 0$. Cette solution représente la formation d'un trou, et la déformation tri-dimensionnelle correspondante est réellement une solution faible de (5). La démonstration de ces faits sera publiée ailleurs. On trouve dans Giusti & Miranda [8] un exemple d'une solution discontinue pour un système non linéaire fortement elliptique, mais leur exemple ne s'applique pas à (5).

Remerciements

Le travail décrit ici a été subventionné par le C.N.R.S. Je veux remercier tous mes amis du Laboratoire d'Analyse Numérique et Fonctionnelle de l'Université Pierre et Marie Curie pour leur intérêt et leur aimable hospitalité.

Bibliographie

1 S.S. Antman, Ordinary differential equations of nonlinear elasticity II : Existence and regularity theory for conservative boundary value problems, Arch. Rat. Mech. Anal. 61 (1976) 353-393.

2 S.S. Antman & H. Brezis, The existence of orientation - preserving deformations in nonlinear elasticity, in "Nonlinear analysis and mechanics : Heriot-Watt Symposium Vol II", ed. R.J. Knops, Pitman, London, 1978.

3 G. Aubert & R. Tahraoui, Théorèmes d'existence pour des problèmes du calcul des variations du type $\text{Inf} \int_0^L f(x,u'(x))dx$ et $\text{Inf} \int_0^L f(x,u(x),u'(x))dx$, J. Differential Eqns, 33 (1979), 1-15.

4 J.M. Ball, Convexity conditions and existence theorems in nonlinear elasticity, Arch. Rat. Mech. Anal. 63 (1977) 337-403.

5 J.M. Ball, Constitutive inequalities and existence theorems in nonlinear elastostatics, in "Nonlinear analysis and mechanics : Heriot-Watt symposium Vol I", ed. R.J. Knops, Pitman, London, 1977.

6 J.M. Ball, Finite time blow-up in nonlinear problems, in "Nonlinear Evolution Equations" ed. M.G. Crandall, Academic Press, New York, 1978.

7 I. Ekeland & R.Témam, "Analyse convexe et problèmes variationnels", Dunod-Gauthier-Villars, Paris, 1974.

8 E. Giusti & M. Miranda, Un esempio di soluzioni discontinue per un problema di minimo relativo ad un integrale regolare del calcolo delle variazioni, Boll. Unione Mat. Ital. Ser. 4, 1 (1968) 219-226.

9 J. Hadamard, "Leçons sur la propagation des ondes", Hermann, Paris 1903.

10 J.K. Knowles & E. Sternberg, On the failure of ellipticity and the emergence of discontinuous deformation gradients in plane finite elasticity, J. of Elasticity 8 (1978) 329-380.

11 L.C. Young, "Lectures on the calculus of variations", Saunders, Philadelphia, 1969.

12 J.M. Ball, Strict convexity, strong ellipticity, and regularity in the calculus of variations, Math. Proc. Camb. Phil. Soc. 87 (1980) 501-513.

J.M. BALL

Department of Mathematics
Heriot-Watt University

Edinburgh EH14 4AS
ECOSSE

H BEIRÃO DA VEIGA*

On the Euler equations for non-homogeneous ideal incompressible fluids**

INTRODUCTION : We consider the motion of a non-homogeneous ideal incompressible fluid in a bounded connected open subset Ω of $\mathbb{R}^3$. We denote by $v(t,x)$ the velocity field, by $\rho(t,x)$ the mass density and by $\pi(t,x)$ the pressure. The equations of the motion are (see Sédov [16 , chap. IV, §1, p.164])

$$\begin{cases} \rho[\dot{v}+(v.\nabla)v-b] = -\nabla\pi & \text{in } Q_{T_0} \equiv]0,T_0[\times\Omega \\ \operatorname{div} v = 0 & \text{in } Q_{T_0}, \\ v \,.\, n = 0 & \text{in } \Sigma_{T_0} \equiv]0,T_0[\times\Gamma \\ \dot{\rho} + v \,.\, \nabla\rho = 0 & \text{in } Q_{T_0}, \\ \rho_{|t=0} = \rho_0 & \text{in } \Omega, \\ v_{|t=0} = a & \text{in } \Omega, \end{cases} \tag{1}$$

where $n = n(x)$ is the unit outward normal to the boundary Γ of Ω, $b = b(t,x)$ is the external force field, and $a = a(x)$, $\rho_0 = \rho_0(x)$ are the initial velocity field and the initial mass density, respectively. By definition $(v.\nabla)\, w = \sum_i v_i D_i w$, $D_i w = \frac{\partial w}{\partial x_i}$, and $\dot{v} = D_t v$. For the case in which the fluid is homogeneous, i.e. the density ρ_0 (and consequently ρ) is constant, equations (1) have been studied by several authors. For non homogeneous fluids,[1] Marsden [13] has stated the existence of a local solution to problem (1) under the assumption that the external force field $b(t,x)$ is zero. Marsden claims that his proof can be extended to the case in which $b(t,x)$ is divergence free and

(*) The result stated in this lecture was proved in paper [6] in collaboration with A. VALLI.

(**) Lecture at the Laboratoire d'Analyse Numérique, Univ. Paris VI (14.6.1979)

(1) For the motion of fluids with salt diffusion see [20].

tangential to the boundary, i.e. div b = 0 in Q_{T_0} and b.n = 0 on Σ_{T_0}. However, for non-homogeneous fluids a general force field cannot be reduced to this particular case (for homogeneous fluids this can be done by subtracting a gradient). Marsden's proof relies on techniques of Riemannian geometry on infinite dimensional manifolds.

In paper [4] (concerning plane motions) H. Beirão da Veiga and A. Valli extend to the non-homogeneous case the technique used by Wolibner [17] and Kato [8] for the homogeneous one. However, this technique gives the existence of a solution only under the additional assumption

$$\left\| \frac{\nabla \rho_0}{\rho_0} \right\|_\infty \leq K,$$

where K is an a priori given constant depending on Ω. In papers [5] (concerning the two and the three dimensional motions, respectively) the same authors drop this assumption by introducing an essential device, the elliptic equations (5). The system (3)+(4)+(5)+(6), which is equivalent to (1), does not contain $\dot{v}$ (compare with system (4.17) in [4]) and this allows us to drop the referred additional assumption. Finally in paper [6] the referred authors give an easier proof of an existence result (now in Sobolev spaces) without the use of lagrangian variables, and prove also a C^∞-regularity result. This lecture concerns the existence result of paper [6], which is the following one :

THEOREM A : Let Γ be of class C^{k+3} and let $a \in H^{k+2}(\Omega)$, $k \geq 1$, with div a = 0 in Ω and a.n = 0 on Γ, $\rho_0 \in H^{k+2}(\Omega)$ with $\rho_0(x) > 0$ for each $x \in \bar{\Omega}$, and $b \in L^1(0,T_0 ; H^{k+2}(\Omega))$.

Then, there exists $T_1 \in]0,T_0]$, $v \in L^\infty(0,T_1;H^{k+2}(\Omega))$ with

$\dot{v} \in L^1(0,T_1;H^{k+1}(\Omega))$, $\rho \in L^\infty(0,T_1;H^{k+2}(\Omega))$ with

$\dot{\rho} \in L^\infty(0,T_1;H^{k+1}(\Omega))$, $\nabla\pi \in L^1(0,T_1;H^{k+2}(\Omega))$ such that (v,ρ,π) is a solution of (1) in Q_{T_1}.

A uniqueness theorem for problem (1) is proved by D. Graffi in [7]; see also [4].

The analytic case on compact manifolds without boundary was studied by Baouendi-Goulaouic [2]. These authors have proved analogous results also for manifolds with boundary (private communication).

For a mathematical study of non-homogenous <u>viscous</u> incompressible fluids see Kazhikhov [9], Ladyženskaja-Solonnikov [10] and Antocev-Kazhikhov [1]; see also the Lions conference [12]. Obviously the viscous case is quite different and one can expect, as usual, stronger results.

Finally the author has proved an existence result for the motion of a <u>compressible</u> ideal fluid in domains with boundary, in an arbitrary force field. The initial velocity field and density distribution are also arbitrary. The proof, which makes use of paper [3], will appear in a forthcoming paper. See [18], [19].

<u>PROOF OF THEOREM A</u>

<u>NOTATIONS</u> : We denote by $\| \ \|_k$ the norm in the Sobolev space $H^k(\Omega)$, by $\| \ \| = \| \ \|_0$ the norm in $L^2(\Omega)$ and by $\| \ \|_{k,T}$ the norm in $L^\infty(0,T;H^k(\Omega))$. Moreover $\| \ \|_\infty$ denotes the norm in $L^\infty(\Omega)$.

In our notations we do not distinguish among scalars and vectors. For instance if v is a vector field we write $v \in H^k(\Omega)$ instead of $v \in (H^k(\Omega))^3$.

Finnaly c denotes any arbitrary constant depending at most on Ω.

An equivalent system of equation : For the sake of convenience we assume in the proof that Ω is simply - connected, that the boundary Γ is connected and that $b \equiv 0$. Moreover, we consider only the case $k = 1$, since the generalization to the case $k > 1$ follows immediately.

First of all recall that under our assumptions a vector field V vanishes in Ω if and only if rot $V = 0$ and div $V = 0$ in Ω, and $V . n = 0$ on Γ. Hence the equation $\dot{v} + (v.\nabla)v = - \frac{\nabla\pi}{\rho}$ is equivalent to the system

$$\begin{cases} \zeta = \text{rot } v & \text{in } Q_{T_0}, \\ \dot{\zeta} + (v.\nabla)\zeta - (\zeta.\nabla)v = - (\nabla\pi) \times \dfrac{\nabla\rho}{\rho^2} & \text{in } Q_{T_0}, \\ \sum_{i,j} (D_i v_j)(D_j v_i) = \dfrac{1}{\rho} \text{div}(-\nabla\pi) - \dfrac{\nabla\rho}{\rho^2}.(-\nabla\pi) & \text{in } Q_{T_0}, \\ - \sum_{i,j} (D_i n_j) v_i v_j = \dfrac{1}{\rho} (-\nabla\pi) . n & \text{in } \Sigma_{T_0}. \end{cases} \qquad (2)$$

Recall that rot $(fV) \equiv f$ rot $V - V \times \nabla f$, div$(fV) \equiv f$ div $V + (\nabla f) . V$ and also the well known [2] formulae $[(v.\nabla)v].n = - \sum_{i,j} (D_i n_j) v_i v_j$ on Γ if $v.n = 0$.

It is then clear that the system (1) is equivalent to the following equations (which are ordered and associated in a meaningfull way) :

$$\begin{cases} \text{rot } v = \zeta & \text{in } Q_{T_0}, \\ \text{div } v = 0 & \text{in } Q_{T_0}, \\ v . n = 0 & \text{in } \Sigma_{T_0}. \end{cases} \qquad (3)_\zeta$$

---

[2] The vectors v and $\nabla(v.n)$ being orthogonal (since $v.n = 0$) one has
$0 = \sum_{i,j} v_i D_i (v_j n_j) = [(v.\nabla)v].n + \sum_{i,j} (D_i n_j) v_i v_j$.

$$\begin{cases} \dot{\rho} + v \cdot \nabla\rho = 0 & \text{in } Q_{T_0}, \\ \rho(0) = \rho_0 & \text{in } \Omega, \end{cases} \tag{4}$$

$$\begin{cases} -\Delta\pi + \dfrac{\nabla\rho}{\rho}.\nabla\pi = \rho \sum\limits_{i,j} (D_i v_j)(D_j v_i) & \text{in } Q_{T_0}, \\ \dfrac{\partial\pi}{\partial n} = \rho \sum\limits_{i,j} (D_i n_j) v_i v_j & \text{on } \Sigma_{T_0}. \end{cases} \tag{5}$$

$$\begin{cases} \dot{\zeta} + (v.\nabla)\,\zeta = -\nabla\pi \times \dfrac{\nabla\rho}{\rho^2} + (\zeta.\nabla)v & \text{in } Q_{T_0}, \\ \zeta(0) = \alpha & \text{in } \Omega, \end{cases} \tag{6}$$

where by definition $\alpha \equiv \text{rot } a(x)$.

Plan of the proof : To solve (3), (4), (5), (6) we use the following device : we replace $(3)_\zeta$ by

$$\begin{cases} \text{rot } v = \phi & \text{in } Q_{T_0}, \\ \text{div } v = 0 & \text{in } Q_{T_0}, \\ v \cdot n = 0 & \text{on } \Sigma_{T_0}, \end{cases} \tag*{$(3)_\phi$}$$

where ϕ is given in a suitable set S, and we estimate v, ρ, π and ζ by solving consecutively the system $(3)_\phi$, (4), (5) and (6). All the estimates depend only on the initial ϕ. If we are able to prove the existence of a fixed point in S for the map "$\phi \to \zeta$" the system (3), (4), (5), (6) is solvable. And we prove that this can be done for a sufficiently small positive T.

Proof of the existence theorem : We consider a

$\phi \in L^\infty(0,T;H^2(\Omega)) \cap C([0,T];H^1(\Omega))$ such that for each $t \in [0,T]$

$$\operatorname{div} \phi = 0 \qquad \text{a.e. in } \Omega \ ; \tag{7}$$

$T > 0$ will be fixed at the end of the procedure. We assume that ϕ belongs to the sphere

$$\|\phi\|_{2,T} \leq A \tag{8}$$

where A will be fixed in the sequel (see (18)). It is clear that there exists a unique solution v of the elliptic system $(3)_{\phi}$, $v \in L^{\infty}(0,T;H^{3}(\Omega)) \cap C([0,T];H^{2}(\Omega))$, and that

$$\|v\|_{3,T} \leq c\ A. \tag{9}$$

Since $v \in L^{\infty}(0,T;C^{1}(\bar{\Omega})) \cap C^{0}(\bar{Q}_{T})$ it is easy to construct the solution $\rho(t,x)$ of equation (4) by using the method of characteristics. Moreover the following estimates for the solution hold :

$$\begin{cases} \|\rho\|_{3,T} \leq \|\rho_0\|_3 \quad e^{cAT} \\ \\ \|\dot{\rho}\|_{2,T} \leq c\ A\, \|\rho_0\|_3 \quad e^{cAT} \end{cases} \tag{10}$$

<u>PROOF</u> : Apply $D_{x_j}\ D_{x_k}\ D_{x_\ell}$ to both sides of equation $(4)_1$; multiply by $D_{x_j}\ D_{x_k}\ D_{x_\ell}\ \rho$ and integrate over Ω. One gets with a schematic notation,

$$\frac{1}{2}\frac{d}{dt}\int_{\Omega} (D_{x_j} D_{x_k} D_{x_\ell} \rho)^2 dx + \frac{1}{2}\sum_i \int_{\Omega} v_i D_i (D_{x_j} D_{x_k} D_{x_\ell} \rho)^2 dx \leq$$

$$\leq 3\int_{\Omega} |Dv|\ |D^3\rho|\ |D_{x_j} D_{x_k} D_{x_\ell} \rho|\ dx + \ldots + \int_{\Omega} |D^3 v|\ |D\rho|\ |D_{x_j} D_{x_k} D_{x_\ell} \rho|\ dx.$$

Integrating by parts the second integral on the left, one sees that it vanishes since $\operatorname{div} v = 0$ in Ω and $v.n = 0$ on Γ. Hence by using Sobolev's imbedding theorems

$$H^2(\Omega) \subset L^\infty(\Omega), \; H^1(\Omega) \subset L^4(\Omega)$$

and Hölder inequalities one obtains easily

$$\frac{1}{2}\frac{d}{dt}\|D_{x_j}D_{x_k}D_{x_\ell}\rho\|^2 \le c\,\|D_{x_j}D_{x_k}D_{x_\ell}\rho\|\,\|v\|_3\,\|\rho\|_3 .$$

Simplifying the term $\|D_{x_j}D_{x_k}D_{x_\ell}\rho\|$ and adding in j,k,ℓ (add also the analogous inequalities obtained for the smaller order derivatives of ρ) it follows that

$$\frac{d}{dt}\|\rho(t)\|_3 \le c\|v\|_{3,T}\,\|\rho(t)\|_3 .$$

The Gronwall's lemma gives now $(10)_1$. The estimate $(10)_2$ follows directly from the equation $(4)_1$ and from the preceeding estimates. □

Now we solve the elliptic equation (5). Since the time is considered as a parameter we write (5) for each fixed t, i.e.

$$\begin{cases} -\Delta\pi + \dfrac{\nabla\rho}{\rho}\cdot\nabla\pi = \rho\sum\limits_{i,j}(D_i v_j)(D_j v_i) \equiv f & \text{in } \Omega, \\ \dfrac{\partial\pi}{\partial n} = \rho\sum\limits_{i,j}(D_i n_j)v_i v_j \equiv g & \text{on } \Gamma. \end{cases} \tag{11}$$

The corresponding homogeneous equations, which can be written in the form

$$\begin{cases} \operatorname{div}(\frac{1}{\rho}\nabla\pi) = 0 & \text{in } \Omega, \\ (\frac{1}{\rho}\nabla\pi)\cdot n = 0 & \text{on } \Gamma, \end{cases} \tag{12}$$

has only the constants as solutions (multiplying $(12)_1$ by π and integrating over Ω one gets easily $\nabla\pi = 0$). Hence also the adjoint homogeneous problem

$$\begin{cases} \Delta\pi^* + \operatorname{div}(\frac{\nabla\rho}{\rho}\pi^*) = 0 & \text{in } \Omega, \\ \frac{\partial\pi^*}{\partial n} + (\frac{\nabla\rho}{\rho} \cdot n)\pi^* = 0 & \text{on } \Gamma \end{cases} \tag{13}$$

has a unique linear independant solution.

By direct computation one verifies that this solution is $\frac{1}{\rho}$. Hence the compatibility condition among the data f and g in equation (11) is

$$\int_\Omega \frac{f}{\rho}\, dx = \int_\Gamma -\frac{g}{\rho}\, d\Gamma,$$

and this condition holds as one verifies easily. Thus (11) is uniquely solvable up to an arbitrary additive constant, hence (5) is uniquely solvable up to an arbitrary additive function of t (for the above argument see C. Miranda [15], theorem 22.I and 22.III, p.84). Moreover from classical estimates for elliptic equations (see C. Miranda [14] theor. 5.1 ; see also Ladyženskaja-Ural'ceva [11]) it follows in particular that

$$\|\nabla\pi\|_3 \le c(\Omega, \|\frac{\nabla\rho}{\rho}\|_2)\ (\|f\|_2 + \|g\|_3).$$

By using (9) and $(10)_1$ one obtains

$$\|\nabla\pi\|_{3,T} \le \bar{c}(A,T) \tag{14}$$

where $\bar{c}$ is non-decreasing as a function of A and T ($\bar{c}$ depends also on the fixed data ρ_0 and on Ω).

Finally,we consider the linear equation (6). The solution can be constructed by using the characteristics method. Moreover, $\zeta \in L^\infty(0,T;H^2(\Omega))$, $\dot{\zeta} \in L^\infty(0,T;H^1(\Omega))$ and

$$\begin{cases} \|\zeta\|_{2,T} \leq [\|\alpha\|_2 + \bar{c}_0(A,T)]\, e^{cAT}, \\ \|\dot{\zeta}\|_{1,T} \leq \bar{c}_1(A,T)\,(1 + \|\alpha\|_2), \end{cases} \tag{15}$$

where $\bar{c}_0$ and $\bar{c}_1$ are non-decreasing functions in the variables A and T (they depend also on Ω and ρ_0) and moreover $\lim_{T\to 0^+} \bar{c}_0(A,T) = 0$.

PROOF : Apply $D_{x_j} D_{x_k}$ to both sides of $\dot{\zeta} + (v.\nabla)\zeta = \nabla\pi \times \nabla(\frac{1}{\rho}) + (\zeta.\nabla)v$; multiply by $D_{x_j} D_{x_k} \zeta$ and integrate over Ω. Arguing as in the proof of $(10)_1$ one gets easily

$$\frac{d}{dt}\|\zeta(t)\|_2 \leq c\,\|v\|_{3,T}\, \|\zeta(t)\|_2 + c\|\nabla\pi\|_{2,T}\, \|\frac{1}{\rho}\|_{3,T}.$$

Recalling in particular that $\|\frac{1}{\rho}\|_{3,T} \leq c(\rho_0)\, \|\rho\|_{3,T}$, since the solution of equation (4) verifies $\min_{\bar{Q}_T} \rho(t,x) = \min_{\bar{\Omega}} \rho_0(x)$, using the Gronwall's lemma and the estimates (9), $(10)_1$ and (14) we obtain $(15)_1$. Finally $(15)_2$ follows from $(6)_1$, (9), $(10)_1$, (14) and $(15)_1$. □

Finally we will show that for each $t \in [0,T]$

$$\operatorname{div} \zeta = 0 \qquad \text{in } \Omega. \tag{16}$$

PROOF : From the general formulae

$$(v.\nabla)\zeta - (\zeta.\nabla)v = v \operatorname{div} \zeta - \zeta \operatorname{div} v - \operatorname{rot}(v \times \zeta),$$

from div v = 0 and from $(6)_1$ one obtains

$$\begin{aligned} \dot{\zeta} + v \operatorname{div} \zeta &= \operatorname{rot}(v \times \zeta) - \nabla\pi \times \frac{\nabla\rho}{\rho^2} \\ &= \operatorname{rot}(v \times \zeta - \frac{\nabla\pi}{\rho}). \end{aligned} \tag{17}$$

By applying the operator div to both sides of (17) one sees that div ζ verifies the equation

$$\frac{\partial(\text{div } \zeta)}{\partial t} + v \cdot \nabla(\text{div } \zeta) = 0.$$

Since $(\text{div } \zeta)_{|t=0} = \text{div } \alpha = 0$, (16) follows. □

<u>Construction of the fixed point</u> : Fixed A such that

$$A > \|\alpha\|_2, \tag{18}$$

say $A = \|\alpha\|_2 + 1$, and also $T_1 \in]0,T_0]$ such that

$$[\|\alpha\|_2 + \bar{c}_0(A,T_1)]\, e^{c A T_1} < A. \tag{19}$$

This can be done, as follows from the property verified by $\bar{c}_0(A,T)$. Put

$$B \equiv \bar{c}_1\,(A,T_1)\,(1 + \|\alpha\|_2) \tag{20}$$

and define the convex set

$$S \equiv \{\phi \text{ in } Q_{T_1} : \|\phi\|_{2,T_1} \leq A,\ \|\dot{\phi}\|_{1,T_1} \leq B, \text{ and div } \phi = 0\}. \tag{21}$$

S is a compact subset of $X \equiv C([0,T_1] ; H^1(\Omega))$ as follows from the Ascoli-Arzelà's theorem. In fact the elements of S are equi-lipschitz continuous in $[0,T_1]$ with values in $H^1(\Omega)$, and bounded sets in $H^2(\Omega)$ are relatively compact in $H^1(\Omega)$.

Let F be the map "$\phi \to \zeta$" defined by the previous construction. By (15), (16) and the above definitions it follows that $F[S] \subset S$.

Moreover $F : S \to X$ is continuous in the X topology.

PROOF : Let ϕ, $\phi_n \in S$, $\phi_n \to \phi$ in X, and let us distinguish with an index n the functions and the equations corresponding to the data ϕ_n. It is clear that the solutions v_n of the linear elliptic equation $(3)_{\phi_n}$ converge in $C([0,T_1] ; H^2(\Omega))$ to the solution v of $(3)_\phi$. Furthermore $\rho_n \to \rho$ in $L^\infty(0,T_1 ; L^2(\Omega))$ since

$$\frac{1}{2}\frac{d}{dt}\|\rho_n - \rho\|^2 \leq \|\rho_n - \rho\| \|\nabla\rho\|_{L^\infty(Q_{T_1})} \|v_n - v\|_{0,T_1}$$

as follows from (4) and $(4)_n$. Hence by (10), using a compactness argument, it follows that $\rho_n \to \rho$ in $C([0,T_1] ; H^2(\Omega))$.

Using the above results concerning v_n and ρ_n and known estimates for the Neumann boundary problem one obtains from (11) and $(11)_n$ with a straightforward calculation that $\nabla\pi_n \to \nabla\pi$ in $L^\infty(0,T_1 ; H^1(\Omega))$. Recall that the norms $\|v_n\|_{3,T}$ and $\|\rho_n\|_{3,T}$ are equi-bounded.

Finally, by evaluating $\frac{d}{dt}\|\zeta_n - \zeta\|^2$ from equations (6) and $(6)_n$ one easily gets $\zeta_n \to \zeta$ in $C([0,T_1] ; L^2(\Omega))$. By the compactness of $\overline{F(S)}$ this implies that $\zeta_n \to \zeta$ in X. □

Now by the Schauder fixed point theorem the map F has a fixed point $\zeta = \phi$ in S. Hence,the corresponding ζ, v, ρ and π are a solution of (3), (4), (5) and (6) in Q_{T_1}. As we have seen, v, ρ and $-\nabla\pi$ are then a solution of (1) in Q_{T_1}.

Remark : In addition to the statement of theorem A we have shown that $\nabla\pi \in L^\infty(0,T_1 ; H^3(\Omega))$, hence by equation $(1)_1$, $\dot{v} \in L^\infty(0,T_1 ; H^2(\Omega))$. In the general case one gets $\nabla\pi \in L^\infty(0,T_1 ; H^{k+2}(\Omega))$ and $\dot{v} \in L^\infty(0,T_1 ; H^{k+1}(\Omega))$.

The same result with the same proof holds if instead of $b \equiv 0$ one has $b \in L^{\infty}(0,T_0\ ;\ H^{k+2}(\Omega))$.

If $b \in L^1(0,T_0\ ;\ H^{k+2}(\Omega)) \cap L^{\infty}(0,T_0\ ;\ H^{k+1}(\Omega))$ then

$$\nabla\pi \in L^1(0,T_1\ ;\ H^{k+2}(\Omega)) \cap L^{\infty}(0,T_1\ ;\ H^{k+1}(\Omega))$$

hence $\dot{v} \in L^{\infty}(0,T_1\ ;\ H^{k+1}(\Omega))$ as in the preceeding case.

Finally if $b \in L^1(0,T_0\ ;\ H^{k+2}(\Omega))$ (see the result stated in theorem A) the proof is similar to the previous one (see [6]).

Bibliography

1 S.N. Antocev - A.V. Kazhikhov, "Mathematical study of flows of non homogeneous fluids", Novosibirsk, Lectures at the University, 1973 (russian).

2 M.S. Baouendi - C. Goulaouic, "Solutions analytiques de l'équation d'Euler d'un fluide incompressible", Seminaire Goulaouic - Schwartz, 1976/77, Paris.

3 H. Beirão da Veiga, "On an Euler type equation in hydrodynamics", Ann. Mat. Pura Appl. (to appear).

4 H. Beirão da Veiga - A. Valli, "On the motion of a non-homogeneous ideal incompressible fluid in an external force field", Rend. Sem. Mat. Padova (1979).

5 H. Beirão da Veiga - A. Valli, "On the Euler equations for non-homogeneous fluids", (I) Rend. Sem. Mat. Padova (to appear), (II) J. Math. Anal. Appl., 73 (1980) 338-350.

6 H. Beirão da Veiga - A. Valli, "Existence of C^{∞} solutions of the Euler equations for non-homogeneous fluids", Comm. Partial

Diff. Eq., 5 (1980), 95-107.

7 D. Graffi, "Il teorema di unicità per i fluidi incompressibili, perfetti, eterogenei", Rev. Un. Mat. Argentina 17 (1955), 73-77.

8 T. Kato, "On classical solutions of the two-dimensional non-stationary Euler equation", Arch. Rat. Mech. Anal., 25 (1967), 188-200.

9 A.V. Kazhikhov, "Solvability of the initial and boundary-value problem for the equations of motion of an inhomogeneous viscous incompressible fluid", Soviet Physics Dokl., 19 (1974), 331-332 (previously in Dokl. Akad. Nauk SS SR, 216 (1974), 1008-1010 (russian)).

10 O.A. Ladyženskaja - V.A. Solonnikov, "The unique solvability of an initial-boundary value problem for viscous, incompressible, inhomogeneous fluids", Naučn. Sem. Leningrad, 52 (1975), 52-109, 218-219 (russian).

11 O.A. Ladyženskaja - N.N. Ural'ceva, "Equations aux derivées partielles de type elliptique", Dunod, Paris (1968) (translated from russian).

12 J.L. Lions, "One some questions in boundary value problems of Mathematical Physics", Rio de Janeiro, Lectures at the Univ. Federal, Instituto de Mat., (1977).

13 J.E. Marsden, "Well-posedness of the equations of a non-homogeneous perfect fluid", Comm. Partial Diff. Eq., 1 (1976), 215-230.

14 C. Miranda, "Sul problema misto per le equazioni lineari ellittiche", Ann. Mat. Pura Appl., 39 (1955) 279-303.

15 C. Miranda, "Partial differential equations of elliptic type", second revised edition, Springer Verlag, Berlin - Heidelberg (1970) (first edition 1955).

16 L. Sédov, "Mécanique des milieux continus", vol. I, Edition MIR, Moscow (1975) (translated from the russian).

17 W. Wolibner, "Un théorème sur l'existence du mouvement plan d'un fluide parfait, homogène, incompressible, pendant un temps infiniment long", Math. Z., 37 (1933) 698-726.

18 H. Beirão da Veiga, "Un théorème d'existence dans la dynamique des fluides compressibles", C.R. Acad. Sç. Paris, 289 (1979) 297-299.

19 H. Beirão da Veiga, "On the motion of compressible barotropic perfect fluids", preprint Università di Trento, UTM 62 - February 1980.

20 H. Beirão da Veiga - R. Serapioni - A. Valli, "On the motion of non-homogeneous fluids in presence of diffusion, to appear in J. Math. Analysis and Appl.

H. BEIRÃO DA VEIGA
Istituto di Matematica
Università di Trento
TRENTO - ITALY

STEPHAN HILDEBRANDT

Boundary behavior of minimal surfaces with free boundaries

During the past hundred years a variety of fascinating free boundary value problems for minimal surfaces has been studied. The investigations have been concerned with the problem of finding minimal surfaces the boundary of which, or at least part of their boundary, is left free on connected closed point sets, say, surfaces. In addition, there are interesting studies of systems of minimal surfaces, of obstacle problems, and of minimal surfaces with movable boundary curves of prescribed length. A fairly complete description of the various results available in the literature can be found in [1] , [3] , [11] , [12] , and [13] .

In the present lecture, we shall report on joint work of J.C.C. Nitsche and S. Hildebrandt which will be published in the two forthcoming papers [5] and [6] . We have been concerned with minimal surfaces having free or partially free boundaries on prescribed supporting surfaces. While satisfactory results exist regarding the existence of solutions and regarding the behavior of a solution surface near the fixed arcs of its boundary, the question of the regularity of a solution surface near its free boundary, and of the geometric nature of its trace was only partially solved. In our papers [5] and [6] , we have tried to answer questions of this kind.

In order to give precise statement of our results, we will consider a typical problem.

Let $<\Gamma, S>$ be a configuration in the three-dimensional Euclidean space $\mathbb{R}^3$ consisting of a smooth two-dimensional surface S and of a smooth Jordan arc Γ having its end points P_1 and P_2 on S, but no other points in common with S. We introduce the class $\mathcal{C} = \mathcal{C}(\Gamma,S)$ of all surfaces $X(w) = (x(w),y(w),z(w))$ in $C^0 \cap H_2^1(B,\mathbb{R}^3)$, $w = u+iv$, parametrized over the semi-disk $B = \{w : |w| < 1\}$, which are bounded by $<\Gamma,S>$ in the following sense :

Let X_C and X_I be the L_2-traces of $X \in H_2^1(B,\mathbb{R}^3)$ on $C = \{w : |w| = 1, v \geq 0\}$ and on $I = \{w : |u| < 1, v = 0\}$, respectively. Then, for any surface $X(w)$ in $\mathcal{C}(\Gamma,S)$, we assume that X_C maps C continuously in a weakly monotonic manner onto Γ such that $X_C(-1) = P_1$, $X_C(1) = P_2$, $X_C(i) = P_3$, where P_3 is a fixed third point on Γ different from P_1 and P_2, while $X_I(w) \in S$ almost everywhere on I.

It is well known that the variational problem

$$P(\Gamma,S) : \quad D(X) = \iint_B (|X_u|^2 + |X_v|^2)\,du\,dv \rightarrow \min$$

among all surfaces $X \in \mathcal{C}(\Gamma,S)$

has always at least one solution $X \in \mathcal{C}(\Gamma,S)$. The vector function $X(w)$ is real analytic in B and satisfies there the conditions

$$\Delta X = 0 \tag{1}$$

$$|X_u|^2 = |X_v|^2, \quad X_u . X_v = 0 . \tag{2}$$

That is, every solution of $P(\Gamma,S)$ is a minimal surface which is bounded in a weak sense by the configuration $<\Gamma,S>$. It is also known that $X(w)$ minimizes as well the area functional

$$A(X) = \iint_B |X_u \wedge X_v|\,du\,dv$$

in the class $\mathcal{C}(\Gamma,S)$. Moreover, there exists satisfactory information concerning the boundary behavior of a solution $X(w)$ of $P(\Gamma,S)$ near the fixed boundary; cf. [11], pp. 281-285. In particular, it is of class $C^{s,\alpha}(B \cup \mathring{C},\mathbb{R}^3)$, $s = 1,2,\ldots,$ $0 < \alpha < 1$, if Γ is a regular arc of class

$C^{s,\alpha}$, and if $\overset{\circ}{C}$ is the interior of the arc C. At the free part of the boundary, the investigations have not yet reached their final stage. The first result in this direction is due to H. Lewy [9] . He proved that, for a compact, regular, analytic supporting surface S without boundary, the solution surface can be continued analytically across the free part of its boundary. In 1970, W. Jäger [8] proved that each solution of $P(\Gamma,S)$ is of class $C^{s,\alpha}$, $s \geq 3$, in $B \cup I$, provided that S is a regular surface of class $C^{s,\alpha}$ without boundary satisfying certain other conditions as well. For the present, the best result is the following : Every solution of $P(\Gamma,S)$ belongs to $C^{s,\alpha}(B \cup I,\mathbb{R}^3)$ provided that S is a regular surface of class $C^{s,\alpha}$, $s = 1,2,\ldots,0 < \alpha < 1$, without boundary, and that S satisfies a local chord-arc condition (cf. [12] , [13]).

That is, S can be for instance a sphere, a torus, or a plane, but the theorem does not apply to surfaces S with boundary. A typical example of such a surface is the finite portion of a plane. On the other hand, these are just the examples with which the experimenter is often confronted. In [5] , J.C.C. Nitsche and the present author have found a regularity theorem for solutions of $P(\Gamma,S)$ yielding regularity up to the free boundary even in cases when the boundary of S is non-void. For this, let us suppose that S is part of a larger complete surface $\mathcal{T}$ without boundary, which is obtained from $\mathcal{T}$ by finitely many cuts along closed and mutually non-intersecting Jordan curves $\Gamma_1,\ldots,\Gamma_N$.

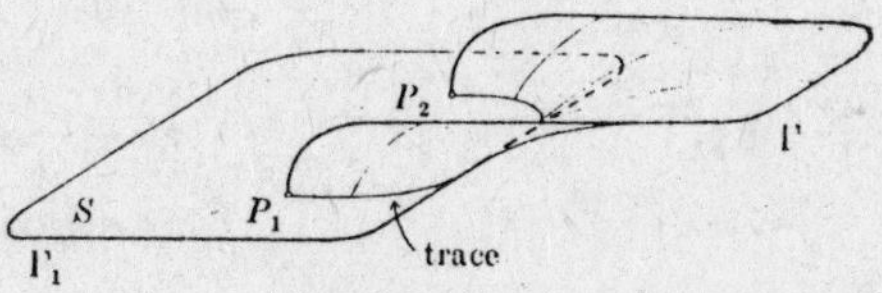

Fig. 1.

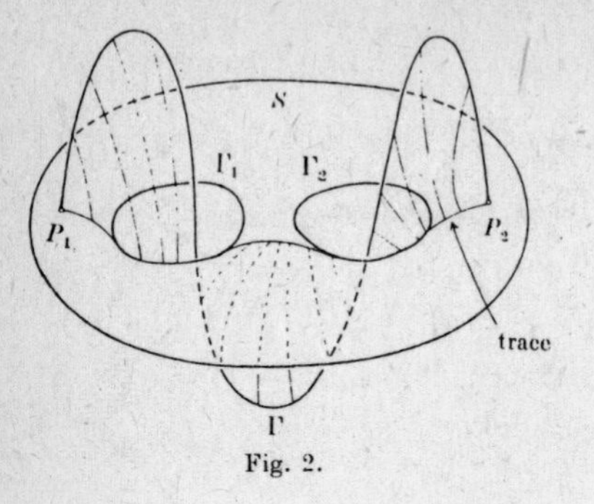

Fig. 2.

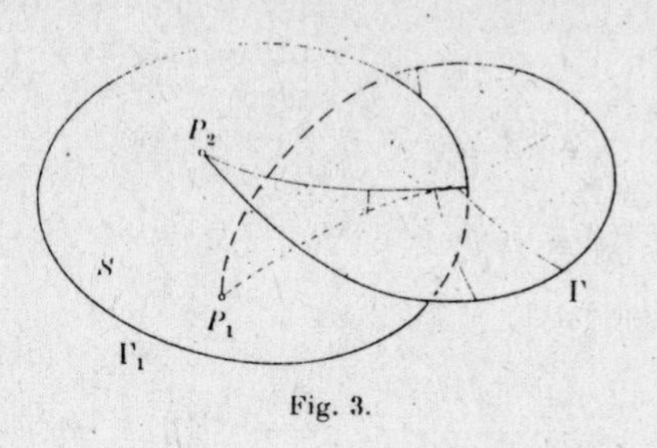

Fig. 3.

Then, the approach of [5] consists in treating $P(\Gamma,S)$ as a Signorini problem, that is, as as variational problem with a "thin obstacle" on the supporting surface $\mathcal{T}$, the obstacle being formed by the curves $\Gamma_1,\ldots,\Gamma_N$. C.Gerhardt has recently treated this problem for non-parametric minimal surfaces. J. Frehse [2] , [3] has recently proved scalar-valued solutions of Signorini problems are in fact of class C^1 up to the thin obstacle. However, these results do not apply to the problem considered in this lecture since we deal with parametric minimal surfaces, that is, with systems of differential equations.

Let us formulate the main result of [5] as

<u>THEOREM 1</u>. : Suppose that $X = X(w)$ is a solution of the variational problem $P(\Gamma,S)$ where S satisfies a local chord-arc condition, and is part of a regular C^3-surface $\mathcal{T}$ in $\mathbb{R}^3$ without boundary which is cut out of $\mathcal{T}$ by finitely many, closed, regular, non intersecting Jordan curves of class C^3. Then, $X(w)$ belongs to the regularity class $C^1(B \cup I, \mathbb{R}^3)$. Moreover, let $w_o \in B \cup I$ be a branch point of $X(w)$, i.e., $X_u(w_o) = X_v(w_o) = 0$. Then, there exists a vector $b = (b^1,b^2,b^3) \in \mathbb{C}^3$ with $|b| \neq 0$ and $b.b = 0$,

and an integer $m \geq 1$, such that

$$X_w(w) = b(w-w_0)^m + o(|w-w_0|^m) \quad \text{as} \quad w \to w_0. \tag{3}$$

Let $b = \frac{1}{2}(\alpha - i\beta)$, $\alpha, \beta \in \mathbb{R}^3$. Then $|\alpha| = |\beta| \neq 0$, $\alpha.\beta = 0$, and

$$X_u(w) \wedge X_v(w) = (\alpha \wedge \beta)\ |w-w_0|^{2m} + o(|w-w_0|^{2m}).$$

Therefore, the normal vector $N(w) = X_u(w) \wedge X_v(w) \ |X_u(w) \wedge X_v(w)|^{-1}$, $w \neq w_0$, tends to the limit vector $\alpha \wedge \beta \ |\alpha \wedge \beta|^{-1}$ as $w \to w_0$. That is, the minimal surface $X(w)$ possesses a tangent plane in each branch point $w_0 \in B \cup I$.

Moreover, if $w_0 = u_0 \in I$ is a boundary branch point, we obtain for the tangent vector $T(u) = X_u(u) \ |X_u(u)|^{-1}$, $u \in I$, the asymptotic representation

$$T(u) = \frac{\alpha}{|\alpha|} \left[\frac{u-u_0}{|u-u_0|}\right]^m + o(1) \qquad \text{as} \quad u \to u_0. \tag{4}$$

Therefore, the non-oriented tangent moves continuously through a boundary branch point while the oriented tangent is continuous for branch points of even order, but, for branch points of odd order, the tangent direction jumps by 180 degrees.

<u>Sketch of the proof</u> : Firstly, we prove that a solution $X(w)$ of $P(\Gamma,S)$ satisfies a Morrey condition

$$\iint\limits_{B \cap B_r(w_0)} |\nabla X|^2 \, du\, dv \leq \left(\frac{2r}{d}\right)^{2\mu} D(X) \tag{5}$$

where $d \in (0,1)$, for all $w_0 \in \{w \in B : |w| < 1-d\}$, and all $r \in (0,\infty)$. The number $\mu > 0$ depends only on the configuration $< \Gamma, S >$. The relation (5) implies that $X(w)$ is of class $C^{0,\mu}$ on each subset $\overline{Z}_d$ of $\overline{B}$, $0 < d < 1$, where $Z_d = \{w \in B : |w| < 1-d\}$. The proof of (5) is given by a modification of the well known reasoning due to Morrey [10], and one uses that the Dirichlet integral is conformally invariant, and that S satisfies a local chord-arc condition. That is, there exist two numbers $\delta > 0$ and $M \geq 1$ with the following property : If X_1 and X_2 are two points on S with $|X_1-X_2| \leq \delta$, then there is a rectifiable arc γ on S connecting X_1 and X_2, and whose length $\ell(\gamma)$ is bounded by

$$\ell(\gamma) \leq M|X_1-X_2| .$$

Secondly, we prove that the second derivatives $\nabla^2 X(w)$ of $X(w)$ are square integrable on each subset Z_d, $0 < d < 1$. In virtue of step 1, and of the fact that X is real analytic in B, it is clearly sufficient to prove that for every $u_0 \in I$, there is a number $r_0 > 0$ such that $\nabla^2 X(w)$ is of class L_2 on $S_{r_0}(u_0) = B \cap B_{r_0}(u_0)$, and, furthermore, that the following discussion can be carried out locally, that is, around small pieces of S which can be flattened. If $X(u_0)$ is an interior point of S, the assertion follows from well known results quoted in [12]. Hence it suffices to consider the case where $X(u_0) \in \Gamma_1 \cup \Gamma_2 \cup \ldots \cup \Gamma_N$. Then, we linearize the boundary conditions by flattening a sufficiently small piece S' containing $X(u_0)$, and by straightening a small piece Γ_k' of the obstacle curve Γ_k which contains $X(u_0)$. This is done by introducing new coordinates $y = g(x)$ mapping S' into the half plane $\{y^1 \geq 0,\ y^2 = 0\}$, and Γ_k' into the y^3 - axis such that $g(X(u_0)) = 0$. The line element

$$ds^2 = dx^j\, dx^j$$

is transformed into

$$ds^2 = g_{jk}(y)\ dy^j\ dy^k\ .$$

The transformed surface $Y(w) = g(X(w))$ satisfies another minimum property which implies that

$$\iint_B g_{jk}(Y)\ \ [Y_u^j\ \phi_u^k + Y_v^j\ \phi_v^k]\ \ du\ dv$$
$$\leq\ -\frac{1}{2}\iint_B \frac{\partial g_{jk}}{\partial y^\ell}(Y)\ \ \{Y_u^j\ Y_u^k + Y_v^j\ Y_v^k\}\ \phi^\ell\ du\ dv \tag{6}$$

for all $\phi \in H_2^1 \cap L_\infty(B,\mathbb{R}^3)$ such that $g^{-1}(Y-\varepsilon\phi)$ lies in $\mathcal{C}(\Gamma,S)$, for $0 < \varepsilon < \varepsilon_o$, $\varepsilon_o > 0$ sufficiently small. It turns out that

$$\phi = -\Delta_{-h}\ \{\eta^2\ \Delta_h\ Y\}$$

is an admissible test function for (6), if η denotes a cut-off function around u_o, with sufficiently small support, and if Δ_h denotes a tangential difference quotient of step width h, and $-\Delta_{-h}$ its adjoint. Employing (5), we are able to bound

$$\iint \eta^2\ \ |\nabla Y|^2\ \ |\Delta_h Y|^2\ du\ dv$$

and

$$\iint \eta^2\ \ |\nabla\ \Delta_h Y|^2\ du\ dv$$

independently of $h \to 0$, whence $|\nabla Y|^2\ |D_u Y|^2$ and $|\nabla D_u Y|^2$ are integrable on $S_{r_o}(u_o)$, $r_o > 0$ sufficiently small.

Furthermore, we infer from (6) that, in $S_{r_o}(u_o)$,

$$\Delta Y^{\ell} + \Gamma^{\ell}_{jk}(Y) \ \{Y^j_u Y^k_u + Y^j_v Y^k_v\} = 0 \tag{7}$$

where Γ^{ℓ}_{jk} are the Christoffel symbols belonging to g_{jk}. Combining these relations with the previous estimates, we finally obtain that $|\nabla Y|^4$ and $|\nabla^2 Y|^2$ are integrable on $S_{r_o}(u_o)$ whence also $|\nabla X|^4$ and $|\nabla^2 X|^2 \in L_1(S_{r_o}(u_o))$. Employing Sobolev's imbedding theorem for functions of two variables we arrive at $X \in H^2_2 \cap H^1_p(Z_d, \mathbb{R}^3)$ for each $d < 1$, and for every $p \in [1,\infty)$. Moreover X_u and X_v have an L_2-trace on every compact subinterval of I. Let $I_{r_o} = \{u \in I : |u-u_o| < r_o\}$. Then, we derive from (6) also the boundary conditions

$$\begin{aligned} g_{j1}(Y)\, Y^j_v &= 0 \quad \text{a.e. on } I_{r_o} \cap \{Y^1(u) > 0\} \\ g_{j3}(Y)\, Y^j_v &= 0 \quad \text{a.e. on } I_{r_o} \\ Y^2 &= 0 \quad \text{on } I_{r_o} \end{aligned} \tag{8}$$

Thirdly, choosing y^1 and y^3 as orthogonal coordinates on S, and letting the y^2-lines intersect S orthogonally, we obtain that $(g_{jk}(y))$ is of the form

$$\begin{pmatrix} E & 0 & 0 \\ 0 & 1 & 0 \\ 0 & 0 & G \end{pmatrix}$$

on S. Therefore, we infer from (8) that

$$Y^1_v = 0 \quad \text{a.e. on } I_{r_o} \cap \{Y^1(u) > 0\}$$

$$Y^3_v = 0 \quad \text{a.e. on } I_{r_o} \tag{9}$$

$$Y^2 = 0 \quad \text{on } I_{r_o}$$

Thus, for $1 < p < \infty$,

$$\Delta Y^2 \in L_p \text{ on } S_{r_o}(u_o) \;, \quad Y^2 = 0 \quad \text{on } I_{r_o} \;,$$

$$\Delta Y^3 \in L_p \text{ on } S_{r_o}(u_o) \;, \quad Y^3_v = 0 \quad \text{a.e. on } I_{r_o} \;.$$

Well known results from potential theory imply that $Y^2(w)$ and $Y^3(w)$ are of class H^2_p on $S_{r_1}(u_o)$, $0 < r_1 < r_o$, and, therefore, also of class $C^{1,\alpha}$ on $\overline{S_{r_2}(u_o)}$, $0 < r_2 < r_1 < r_o$, for each $\alpha \in (0,1)$.

Transforming the conformality relations (2) to the new variables, we can express $(Y^1_w)^2$ in terms of Y^2_w, Y^3_w, and of $g_{jk}(Y)$, $\frac{\partial}{\partial w} = \frac{1}{2}(\frac{\partial}{\partial u} - i\,\frac{\partial}{\partial v})$. This enables us to conclude that Y^1_w is continuous on $S_{r_2}(u_o)$.

Thus we arrive at $X \in C^1(B \cup I, \mathbb{R}^3)$. In the last step of our reasoning, we establish the expansion (3). Employing once more the conformality relations (3), we see that

$$|\Delta Y^2| + |\Delta Y^3| \leq c \;\{|\nabla Y^2|^2 + |\nabla Y^3|^2\} \tag{10}$$

in $S_{r_2}(u_o)$, and that $Y^2(w)$, $Y^3(w)$ are of class $C^{1,\alpha}$ on $S_{r_2}(u_o)$, and satisfy $Y^2(u) = 0$ and $Y^3_v(u) = 0$ for $u \in I$ with $|u-u_o| \leq r_1$.

Extending Y^2 and Y^3 across I by the definitions $Y^2(\overline{w}) = Y^2(w)$, $Y^3(\overline{w}) = -\,Y^3(w)$, we see that Y^2 and Y^3 are of class C^1 on $\overline{B_{r_2}(u_o)}$.

On the other hand, (10) and a well known application of the Gauss integration formula yield for $\widetilde{Y} = (Y^2, Y^3)$ the formula

$$\left| \oint_{\partial\mathfrak{D}} \tilde{Y}_w \phi \, dw \right| \leq 2 \iint_{\mathfrak{D}} |\tilde{Y}_w| \{|\phi_{\bar{w}}| + k\,|\phi|\} \, du\, dv \qquad (11)$$

for all $\mathfrak{D} \subset B_\rho(w_o)$ with piecewise smooth boundary and for all $\phi \in C^1(\mathfrak{D};\mathbb{C}^2)$, where k is a fixed number. As it has been noticed by E. Heinz, one can now apply the technique of Hartman and Wintner [4] to prove that there is a vector $A = (A^2, A^3) \in \mathbb{C}^2$, $A \neq 0$, and an integer $m \geq 1$, such that

$$\tilde{Y}_w(w) = A(w-w_o)^m + o(|w-w_o|^m) \quad \text{as} \quad w \to w_o, \qquad (12)$$

provided that $w_o \in B_{r_2}(u_o)$ and that $Y_w(w_o) = 0$.

Applying the expansion (12) to a possible boundary branch point u_o, and using once more the conformality relations (2) we obtain the expansion (3) for $X(w)$ around u_o. For interior branch points $w_o \in B$, the expansion (3) is well known; cf. [11].

This concludes the proof of Theorem 1.

Next, we shall describe the results of [6]. This investigation provides a mathematical discussion of a special configuration $<\Gamma, S>$ for which a uniqueness result for the variational problem $P(\Gamma, S)$ can be proved. Then we can describe special situations in which cusps, i.e. branch points of odd order, must appear, as well as other situations where cusps cannot be present. For the following, we assume throughout that the configuration $<\Gamma, S>$ satisfies the

<u>Assumption A.</u> : The supporting surface is the half plane $\{(x,y,z) : x \geq 0, \ y = 0\}$. The curve Γ is a regular arc of class $C^{1,\alpha}$, $0 < \alpha < 1$, which does not meet S, except for its end points P_1 and P_2 where Γ issues from S at right angles. The points P_1 and P_2 are

different from each other and lie on the interior part of S. Moreover, Γ is symmetric with respect to the x-axis, and the orthogonal projection onto the (x,y)-plane maps Γ in a one-to-one way onto a closed, strictly convex, and regular curve γ of class $C^{1,\alpha}$, except for the points P_1 and P_2 which are projected at the same point of γ, Finally, close to P_1, Γ lies above the (x,z)-plane (except for P_1).

We can assume that P_1 is the end point of Γ with the negative z-coordinate. Let $p(s) = (p^1(s), p^2(s), p^3(s))$, $0 \leq s \leq L$, be the parametrization of Γ by its arc length s, such that $P_1 = p(0)$, $P_2 = p(L)$. Furthermore, denote by $P_3 = p(L/2)$ the uniquely determined intersection point of Γ with the x-axis. Then, there exist positive numbers a, b, and c, such that $P_1 = (a,0,-c)$, $P_2 = (a,0,c)$, $P_3 = (-b,0,0)$.

Let $X = X(w)$ ba a solution surface for the minimum problem $P(\Gamma,S)$, and let $\phi = \phi(w)$ be a test function such that $X + \varepsilon\phi$ belongs to the class $\mathcal{C}(\Gamma,S)$ of admissible surfaces, for all $\varepsilon \in [0,\varepsilon_0]$, where $\varepsilon_0 = \varepsilon_0(\phi)$ denotes a positive number which may depend on ϕ. Then,

$$D(X + \varepsilon\phi) \geq D(X),$$

whence

$$\frac{d}{d\varepsilon} D(X + \varepsilon\phi)\Big|_{\varepsilon=0} \geq 0 .$$

That is,

$$D(X,\phi) = \iint_B \nabla X . \nabla\phi \, du \, dv \geq 0 .$$

We shall use this relation to define a stationary minimal surface for the configuration $<\Gamma,S>$.

DEFINITION. : A minimal surface $X = X(w)$ will be called a stationary minimal surface for the configuration $<\Gamma,S>$ if it is continuous on $\overline{B}$, and if it lies in the set $\mathscr{C}(\Gamma,S)$ of admissible surfaces and satisfies

$$D(X,\phi) \geq 0$$

for all test functions $\phi = (\phi^1(w),\phi^2(w),\phi^3(w))$ with the property that $X + \varepsilon\phi \in \mathscr{C}(\Gamma,S)$ for $0 \leq \varepsilon < \varepsilon_o$, where $\varepsilon_o = \varepsilon_o(\phi)$ denotes some positive number.

Moreover, we denote by $P^*(\Gamma,S)$ the problem to find a stationary minimal surface for $<\Gamma,S>$.

By the same technique as in the proof of Theorem 1 (cf. step 1), we can show that a solution $X(w)$ of $P(\Gamma,S)$ is continuous in the corners $w = \pm 1$. Therefore, we get

LEMMA 1. : A solution of the minimum problem $P(\Gamma,S)$ is also a solution of the stationary problem $P^*(\Gamma,S)$.

LEMMA 2. : Let $X(w) = (x(w),y(w),z(w))$ be a solution of $P^*(\Gamma,S)$. Then $X(w)$ is of class $C^1(\overline{B},\mathbb{R}^3)$, and

$$y(u) = 0, \quad z_v(u) = 0 \quad \text{for} \quad u \in I. \tag{13}$$

The set

$$I_1 = \{u \in I : x(u) > 0\}$$

is open and contains $(-1,-1+2\delta_o) \cup (1-2\delta_o,1)$ for some $\delta_o > 0$, and

$$x_v(u) = 0 \quad \text{for all} \quad u \in I_1 . \tag{14}$$

The set of contact

$$I_2 = \{u \in I : x(u) = 0\}$$

is closed.

The <u>proof</u> uses only arguments which have already been applied for the proof of Theorem 1.

Note that the surface $X(w)$ intersects S orthogonally along $X(I_1)$ when $X(w)$ is a solution of $P^*(\Gamma,S)$.

A simple application of the maximum principle and of E. Hopf's lemma yields

$$-b < x(w) < a \quad \text{for all} \quad w \in B \tag{15}$$

as well as

$$0 \leq x_o = \min\{x(u) : u \in I\} < a . \tag{16}$$

The function $X(w)$ maps the circular arc C homeomorphically onto Γ.

Hence, by assumption A, we obtain for each value $\mu \in (-b,a)$ exactly two points $w_1(\mu) = e^{i\theta_1(\mu)}$ and $w_2(\mu) = e^{i\theta_2(\mu)}$ on C, $0 < \theta_1(\mu) < \theta_2(\mu) < \pi$ such that $x(w_1(\mu)) = x(w_2(\mu)) = \mu$. In addition, we set $w_1(-b) = w_2(-b) = i$, $\theta_1(-b) = \theta_2(-b) = \frac{\pi}{2}$, $w_1(a) = 1$, $w_2(a) = -1$, $\theta_1(a) = 0$, $\theta_2(a) = \pi$.

For each value of $\mu \in \mathbb{R}$, we define the following open subsets of B :

$$\begin{aligned} B^+(\mu) &= \{w \in B : x(w) > \mu\} \\ B^-(\mu) &= \{w \in B : x(w) < \mu\} \\ B(\mu) &= \{w \in B : x(w) \neq \mu\} . \end{aligned} \tag{17}$$

In virtue of (15), we infer that

$$B^+(\mu) = \emptyset \quad , \quad B^-(\mu) = B \quad \text{if} \quad \mu \geq a$$
$$B^+(\mu) = B \quad , \quad B^-(\mu) = \emptyset \quad \text{if} \quad \mu \leq -b \, . \tag{18}$$

<u>LEMMA 3</u>. : For each $\mu \in (-b,a)$, the set $B^-(\mu)$ is connected, and the set $B^+(\mu)$ can have at most two components.

The proof rests on a suitable combination of (14), E. Hopf's lemma and of the maximum principle.

<u>THEOREM 2</u>. : Let $X(w) = (x(w), y(w), z(w))$ be a solution of the stationary problem $P^*(\Gamma,S)$, and let

$$I_1 = \{u \in I : x(u) > 0\} \quad , \quad I_2 = \{u \in I : x(u) = 0\} \, ,$$

$$x_o = \min \{x(u) : u \in I\} \, .$$

Then only the following three cases are possible :

I) $x_o = 0$, and I_2 consists of a single point;

II) $x_o = 0$, and I_2 is a closed interval of positive length;

III) $x_o > 0$, i.e. , I_2 is empty, and there exists a unique value $u_o \in I$ such that $x_o = x(u_o)$, and that $x(u) > x_o$ for $u \neq u_o$, $u \in I$.

<u>Proof</u> : Consider the connected set $B^-(x_o)$. By definition of x_o , the intersection of the boundary of $B^-(x_o)$, with the interval I is nonempty. Let $u_o = \min \{u \in I \cap \partial B^-(x_o)\}$, $u_o' = \max \{u \in I \cap \partial B^-(x_o)\}$.

Then we distinguish the following mutually disjoint cases

(α) $u_o = u'_o$;

(β) $u_o < u'_o$, $x(u) = x_o$ for all $u \in [u_o, u'_o]$;

(γ) $u_o < u'_o$, $x(\tilde{u}) > x_o$ for some $\tilde{u} \in [u_o, u'_o]$.

We claim that the case (γ) cannot happen. For this purpose, we note that, in case (γ), we can find two points u_1 , $u_2 \in [u_o, u'_o]$, $u_1 < u_2$, such that $x(u_1) = x(u_2) = x_o$, and that $x(u) > x_o$ for $u_1 < u < u_2$. Set $m = \max\{x(u) : u_1 < u < u_2\}$, $0 \leq x_o < m < a$, and assume that $x(u') = m$, $u_1 < u' < u_2$. Then $x_u(u') = 0$ as well as $x_v(u) = 0$ for $u_1 < u < u_2$, on account of Lemma 2. Therefore, $x(w)$ can be continued analytically as a harmonic function across the segment $u_1 < u < u_2$ of the u-axis into the lower half plane. Since $\nabla x(u^1) = 0$, we must have an expansion

$$x(w) = m + \text{Re}\{\chi(w-u')^\nu + \ldots\} ,$$

$\chi \neq 0$, $\nu \geq 2$, in a neighborhood of $w = u'$. Since $u = u'$ is a strict local maximum of $x(u)$ on I, we have $\chi < 0$, and $\nu = 2n$, $n \geq 1$. A neighborhood of $w = u'$ in B is divided into $2n+1$ - at least three - open sectors σ_1 , σ_2 ,..., σ_{2n+1} , such that $x(w) < m$ in σ_1 , σ_3,..,σ_{2n+1} and that $x(w) > m$ in σ_2 , σ_4 ,..., σ_{2n} . Consider the set $B(m)$, and denote by Q_1 , Q_2 ,..., Q_{2n+1} those of its components which contain the sectors σ_1 , σ_2 ,..., σ_{2n+1} , respectively. For topological reasons, and because of the maximum principle, the open sets. Q_j are mutually disjoint. Therefore, not both components Q_1 and Q_{2n+1} can be identical with the set $B^-(m)$, and we have a contradiction to Lemma 3.

Thus, the case (γ) is ruled out. Next, we show that (β) is impossible if $x_o > 0$, since Lemma 2 yields $x_v(u) = 0$ for $u_o \leq u \leq u'$, whence $x(w) \equiv x_o$ on $\overline{B}$, because of (β), taking the principle of unique continuation into account. Since $x(w)$ is non constant on $\overline{B}$, we conclude that only the case (α) can appear if $x_o > 0$, and the Theorem is proved.

The case III cannot be excluded for stationary solutions as one can see by considering explicit examples. However, in Theorem 4, we shall rule out this case for solutions of the minimum problem $P(\Gamma,S)$.
The classification of the stationary solutions given in Theorem 2 will be fundamental for our further discussion. Note also that we always suppose assumption A to be satisfed. We begin with three uniqueness results.

THEOREM 3. : Let $X_1(w)$ and $X_2(w)$ be two solutions of the stationary problem $P^*(\Gamma,S)$ which are not of type III. Then,

$$X_1(w) \equiv X_2(w) \quad \text{on} \quad \overline{B} .$$

As we have noted before, a stationary solution for $<\Gamma,S>$ can very well have a trace on I which does not touch the z-axis. For minima, however, the situation is different.

THEOREM 4. : A solution of the minimum problem $P(\Gamma,S)$ cannot be of type III.

From these two theorems, we obtain the following main uniqueness result.

THEOREM 5. : Let $\mathcal{G}(\Gamma,S)$ be the subclass of those solutions of the stationary problem $P^*(\Gamma,S)$ which are not of type III. Then $\mathcal{G}(\Gamma,S)$ contains exactly one surface which, in addition, is the - uniquely determined - solution of the minimum problem $P(\Gamma,S)$.

Theorem 3 will be derived from the following result providing a non-parametric representation for solutions of $P^*(\Gamma,S)$.

THEOREM 6. : Let $X(w)$ be a solution of $P(\Gamma,S)$, and let $x_o = \min\{x(u) : u \in I\}$. Denote by $D = D(x_o)$ the twodimensional domain in the x, y-plane which is obtained from the interior of the orthogonal projection γ of Γ by slitting it along the x-axis from $x = x_o$ to $x = a$. In defining the boundary ∂D of the slit domain D, both borders of the slit $x_o < x \leq a$ will appear but oppositely oriented.

Then, the functions $x(w)$, $y(w)$ provide a C^1-mapping of $\overline{B}$ onto $D \cup \hat{\partial}D$ which is topological except, in case II, for the interval of coincidence $I_2 = \{u \in I : x(u) = 0\}$ which corresponds wholly to the point $(0,0)$ on $\hat{\partial}D$. Furthermore, the minimal surface $\mathcal{M}$, represented by $X(w)$, admits a non-parametric representation $z = Z(x,y)$ over the domain D. The function $Z(x,y)$ is real analytic in D and on both shores of the open segment $x_o < x < a$, and

$$\lim_{y \to +0} \frac{\partial}{\partial y} Z(x,y) = \lim_{y \to -0} \frac{\partial}{\partial y} Z(x,y) = 0 \quad \text{for } 0 < x < a. \qquad (19)$$

Moreover, $Z(x,y)$ is continuous on $D \cup \hat{\partial}D$ in case I and III. In case II, $Z(x,y)$ is continuous on $D \cup \hat{\partial}D - \{(0,0)\}$, and $Z(x,y)$ remains bounded upon approach of the point $(0,0)$.

Let us see how we can obtain Theorem 3 from Theorem 6. For this purpose, let $X_1(w) = (x^{(1)}(w), y^{(1)}(w), z^{(1)}(w))$ and $X_2(w) = (x^{(2)}(w), y^{(2)}(w), z^{(2)}(w))$ be two solutions of the stationary problem $P^*(\Gamma,S)$ such that

$$x_o = \min\{x^{(1)}(u) : u \in I\} = \min\{x^{(2)}(u) : u \in I\}.$$

Then, the domain D, as defined in Theorem 6, is the same for both surfaces $X_1(w)$ and $X_2(w)$. While the mappings $x = x^{(j)}(w)$, $y = y^{(j)}(w)$ might be different for both surfaces, both transform the semi-disc B into the region D, and both minimal surfaces admit a non-parametric representation

$$z = Z_j(x,y) \;, \quad (x,y) \in D \cup \hat{\partial} D \;, \qquad j = 1,2,$$

with properties as stated in Theorem 6. Moreover,

$$Z_1(x,y) = Z_2(x,y) \quad \text{for} \quad \text{all points } (x,y) \in \Upsilon \,. \qquad (20)$$

We set

$$p_j = \frac{\partial}{\partial x} Z_j \;, \; q_j = \frac{\partial}{\partial y} Z_j \;, \quad W_j = \sqrt{1+p_j^2+q_j^2} \;.$$

For small positive numbers δ and ε, denote by $D_{\delta,\varepsilon}$ the set of all points in D whose distance from the boundary of D exceeds δ, and whose distance from the points $(0,0)$ and $(a,0)$ exceeds ε. Let Q be an arbitrary compact subset of $D_{\delta,\varepsilon}$ and call $m(Q)$ the maximum of $W_1(x,y)$ and $W_2(x,y)$ in Q. An application of § 585 in [11] yields the following inequality :

$$\frac{1}{2m(Q)} \iint_Q \{(p_1-p_2)^2 + (q_1-q_2)^2\} \, dx\, dy$$

$$\leq \int_{\partial D_{\delta,\varepsilon}} (Z_1-Z_2) \;\; [-(\frac{q_1}{W_1} - \frac{q_2}{W_2})dx + (\frac{p_1}{W_1} - \frac{p_2}{W_2})dy] \;.$$

Keep ε fixed, and let δ go to zero. In view of

$$\lim_{y \to \pm 0} \frac{\partial}{\partial y} Z_j(x,y) = 0 \quad \text{for} \quad x_o < x < a \;, \quad j = 1,2, \qquad (21)$$

and of (20), we see that certain parts of the line integral tend to zero, so that

$$\frac{1}{2m(Q)} \iint_Q \{(p_1-p_2)^2 + (q_1-q_2)^2\} \, dx\, dy$$
$$\leq \int_{C_\varepsilon} (Z_1-Z_2) \left[- \left(\frac{q_1}{W_1} - \frac{q_2}{W_2}\right) dx + \left(\frac{p_1}{W_1} - \frac{p_2}{W_2}\right) dy \right]$$

Where C_ε denotes the parts of the circles $\{x^2+y^2 = \varepsilon^2\}$ and $\{(x-a)^2+y^2=\varepsilon^2\}$, which are contained in $D \cup \hat{\partial} D$. Since the functions Z_j as well as the quotients p_j/W_j and q_j/W_j are bounded on D, the line integral $\int_{C_\varepsilon}$ tends to zero as $\varepsilon \to 0$, whence $p_1(x,y) \equiv p_2(x,y)$, $q_1(x,y) = q_2(x,y)$ in Q. Since Q was an arbitrary set in $D_{\delta,\varepsilon}$, we infer that $\nabla Z_1(x,y) \equiv \nabla Z_2(x,y)$ in D whence $Z_1(x,y) \equiv Z_2(x,y)$ in D, on account of (20). Therefore, $X_1(w)$ and $X_2(w)$ are conformal representations of the same non-parametric minimal surface $\mathcal{M}$

$$z = Z(x,y) \,, \quad (x,y) \in D \,,$$

with the same parameter domain B, and satisfying the same three point condition on ∂B. Both parametrizations $X_1(w)$ and $X_2(w)$ map B in an one-to-one way onto $\mathcal{M}$. From this, we may conclude that $X_1(w) \equiv X_2(w)$ because a conformal map of B onto itself has to be the identical map if it leaves three points on ∂B fixed.

Our reasoning can be applied to surfaces $X_1(w)$, $X_2(w)$ of type I or II since they satisfy $x_0 = 0$, and, thus, Theorem 3 is proved.

However, the arguments employed above yield more. In particular, we can conclude that each minimal surface $X(w) = (x(w),y(w),z(w))$ is symmetric with respect to the x-axis. For this purpose, we note that

$$X^*(u+iv) = (x(-u+iv),-y(-u+iv),-z(-u+iv))$$

is also a solution of $P^*(\Gamma,S)$ since Γ is symmetric with respect to the x-axis. Since $\min\{x(u) : u \in I\} = \min\{x(-u) : u \in I\}$, the previous reasoning yields that $X(w) \equiv X^*(w)$ on B.

Moreover, if $z = Z(x,y)$ is the non-parametric representation of the minimal surface $\mathcal{M}$ given by $X(w)$, then $Z(x,y) = -Z(x,-y)$, because of the same argument.

Thus we obtain

<u>THEOREM 7.</u> : Every solution $X(w) = (x(w),y(w),z(w))$ of the stationary problem $P^*(\Gamma,S)$ is symmetric with respect to the x-axis, that is,

$$x(u+iv) = x(-u+iv) \tag{22}$$

$$y(u+iv) = -y(-u+iv) \tag{23}$$

$$z(u+iv) = -z(-u+iv), \tag{24}$$

In case I or III, we have

$$x_o = x(0) , \quad x_o < x(u) < a \quad \text{for } u \in I, \ u \neq 0. \tag{25}$$

In case II, I_2 is of the form $[u_1,u_2]$, where $0 < u_2 < 1$ and $u_1 = -u_2$.

Clearly, the relations (23) and (24) imply

$$y(iv) = z(iv) = 0 \quad \text{for all } v \in [0,1] \tag{26}$$

Finally, the non-parametric representation $z = Z(x,y)$ of the minimal surface $\mathcal{M}$, given by $X(w)$, satisfies

$$Z(x,y) = -Z(x,-y) \quad \text{for } (x,y) \in D(x_o) \tag{27}$$

whence

$$\lim_{y\to+0} Z(x,y) = -\lim_{y\to-0} Z(x,y) \quad , \quad x \neq 0 \ , \ \text{in case II.}$$

In virtue of these results, we can derive the following expansions for a solution $X(w)$ of $P(\Gamma,S)$. We omit the details of the proof.

THEOREM 8. : Let $X(w) = (x(w),y(w),z(w))$ be a solution of $P^*(\Gamma,S)$. If $X(w)$ is of class I or III, then $w = 0$ is a first order branch point of $X(w)$, and we have the expansion

$$\left.\begin{aligned} x(w) &= x_o + \mathrm{Re}\,\{ \kappa w^2 + \ldots\} \\ y(w) &= \mathrm{Re}\,\{ i\kappa w^2 + \ldots\} \\ z(w) &= \mathrm{Re}\,\{\mu w^{2n+1} + \ldots\} \end{aligned}\right\} \text{ as } w \to 0 \, , \qquad (28)$$

where $\kappa > 0$, μ real, $\mu \neq 0$, and n is an integer ≥ 1.

If $X(w)$ is of type II, and if $[u_1,u_2]$ is the set of contact I_2, $-1 < u_1 < 0 < u_2 < 1$, $u_1 = -u_2$, then, there exist positive numbers κ and μ, and a réal number $z_1 \neq 0$, such that

$$\left.\begin{aligned} x(w) &= \mathrm{Re}\,\{ i\kappa(w-u_1)^{3/2} + \ldots\} \\ y(w) &= \mathrm{Re}\,\{-i\mu(w-u_1) + \ldots\} \\ z(w) &= \mathrm{Re}\,\{z_1+(-1)^n \mu(w-u_1) + \ldots\} \end{aligned}\right\} \text{ as } w \to u_1 \, , \qquad (29)$$

and

$$\left.\begin{aligned} x(w) &= \mathrm{Re}\,\{ \kappa(w-u_2)^{3/2} + \ldots\} \\ y(w) &= \mathrm{Re}\,\{ i\mu(w-u_2) + \ldots\} \\ z(w) &= \mathrm{Re}\,\{ -z_1+(-1)^n \mu(w-u_2) + \ldots\} \end{aligned}\right\} \text{ as } w \to u_2 \qquad (30)$$

and where $n = 1$ if $z_1 > 0$, $n = 2$ if $z_1 < 0$.

Moreover, no point on I is a branch point of $X(w)$.

This theorem implies the following description of the geometric nature of the trace $X(u)$.

<u>THEOREM 9.</u> : In the cases I and III, the trace $X(u)$, $u \in I$, of a solution $X(w)$ of the stationary problem $P^*(\Gamma,S)$ is a regular, real analytic curve, except for the point $X(0) = (x_0,0,0)$ where the trace curve possesses a cusp. The parameter representation $X(u)$ of the trace is of class C^1 on I, and $w = 0$ is a branch point of first order of the surface $X(w)$. In case II, $X(w)$ has no branch point on I. The trace $X(u)$, $u \in I$, is a regular curve of class $C^{1,1/2}$.

Finally, we will describe a situation in which no cusps can appear on the trace $X(u)$, $u \in I$.

<u>THEOREM 10.</u> : Suppose that the open subarc of Γ with the end points P_1 and P_3 lies in the half space $\{z < 0\}$, and that the open subarc between P_3 and P_2 is contained in the half space $\{z > 0\}$. Then each solution $X(w)$ of the stationary problem $P^*(\Gamma,S)$ has to be of type II. Therefore, its trace $X(u)$, $u \in I$, on S is a regular curve of class $C^{1,1/2}$, i.e., without cups.

<u>Proof</u> : Let $C^+ = \{e^{i\theta} : 0 < \theta < \pi/2\}$, $C^- = \{e^{i\theta} : \pi/2 < \theta < \pi\}$, and let $X(w) = (x(w),y(w),z(w))$ be the considered solution of $P^*(\Gamma,S)$. Then, $z(w) > 0$ for $w \in C^+$, and $z(w) < 0$ for $w \in C^-$. Denote by Q^+ and Q^- the two components of $\{w \in B : z(w) \neq 0\}$, for which $C^+ \subset \partial Q^+$ and $C^- \subset \partial Q^-$ respectively. We claim that there exist no other components of $\{w \in B : z(w) \neq 0\}$. In fact, if R were such a component different from Q^+ and Q^-, then $\partial R \subset B \cup I \cup \{i\}$. In view of the maximum principle,

$z(w)$ cannot vanish everywhere on ∂R. Hence, there is a point on I where $z(w)$ is different from zero, say, positive. Moreover, $\partial R \cap I$ is contained in a compact subinterval of I. Therefore, there is a point $u' \in I$ such that $z(u') = \max\{z(u) : u \in \partial R \cap I\} = \max\{z(w) : w \in \partial R\} > 0$. Clearly, a whole interval on I around u' is also contained in $\partial R \cap I$. Then, by E. Hopf's lemma, $z_v(u') > 0$ which contradicts the relation $z_v(u) = 0$ holding for all $u \in I$.

Since Q^+ and Q^- are the only components of $\{w \in B : z(w) \neq 0\}$ we conclude that $Q^+ = \{re^{i\theta} : 0 < r < 1 \ , \ 0 < \theta < \pi/2\}$, and that $Q^- = \{re^{i\theta} : 0 < r < 1, \pi/2 < \theta < \pi\}$. In view of the - by now standard- reasoning, there can be no zeros of $z_u(u)$ on $(-1,0)$ or on $(0,1)$. Thus, $z_u(u) = 0$ can only hold for $u = 0$. In the class I or III, expansion (28) implies that a neighborhood of $u = 0$ in B is divided into $2n+2$ - at least four - open sectors $\sigma_1, \sigma_2, \ldots, \sigma_{2n+2}$ such that $z(w) > 0$ in $\sigma_1, \sigma_3, \ldots, \sigma_{2n+1}$, and $z(w) < 0$ in $\sigma_2, \sigma_4, \ldots, \sigma_{2n+2}$. This is clearly impossible. Thus only the case II can be possible, q.e.d.

Now, we come to the

Proof of Theorem 4. : Because of (23), $y(iv)$ vanishes for all $v \in [0,1]$. Since $x(0) = x_o \geq 0$, and $x(iv) = -b < 0$, there exists a least number v_1 in $[0,1)$ such that $x(iv_1) = 0$. Suppose now that $X(w)$ is a solution of $P(\Gamma,S)$ which is of type III. Then, $0 < v_1 < 1$. Denote by B' the slit domain obtained by cutting the semi-disk B along the imaginary axis from $w = 0$ til $w = iv_1$. Furthermore, let $w = \tau(\zeta)$ be the conformal mapping from B onto B' leaving the three points $w = +1, -1, i$ fixed. Then, $Y(\zeta) = X(\tau(\zeta))$ is again of class $\mathcal{C}(\Gamma,S)$, since $y(iv) = 0$ for all $v \in [0,1]$. From the invariance of the Dirichlet integral with respect to conformal mappings we conclude that $Y(\zeta)$ is also a solution of the minimum problem $P(\Gamma,S)$ but of type I, in virtue of Theorem 1.

By (28), $Y(\zeta) = (y^1(\zeta), y^2(\zeta), y^3(\zeta))$ possesses an expansion near $\zeta = 0$ of the form

$$y^1(\zeta) = \mathrm{Re}\,\{\kappa\zeta^2 + \dots\}$$
$$y^2(\zeta) = \mathrm{Re}\,\{i\kappa\zeta^2 + \dots\} \qquad (31)$$
$$y^3(\zeta) = \mathrm{Re}\,\{\mu\zeta^{2n+1} + \dots\}.$$

Where $\kappa > 0$, $\mu > 0$ or < 0, $n \geq 1$. Let $\zeta = \alpha + i\beta$. We infer from (31) that the images of segments $(-\varepsilon, 0)$ and $(0, \varepsilon)$, $\varepsilon > 0$, on I under $Y(\alpha)$ are different, i.e., $y^3(-\alpha) \neq y^3(\alpha')$ if $0 < \alpha, \alpha' < \varepsilon$.

On the other hand, the relation $z(iv) = 0$ for $0 \leq v \leq 1$ implies that $y^3(\alpha) = 0$ for $0 \leq |\alpha| \leq \varepsilon$, $\alpha \in I$, if ε is a sufficiently small, positive number. Such a discrepancy is not possible, and $X(w)$ cannot be of type III. Hence, Theorem 4 is proved.

Finally, we have to prove Theorem 6 which provides us with the non-parametric representation. Unfortunately, its proof is too lengthy to reproduce it here. Besides topological arguments and applications of the maximum principle, our reasoning is essentially based on a well known lemma due to Radó (cf. [11], § 373). For the details, we refer to [6].

One checks easily that, for sufficiently small values of the parameter $\lambda > 0$, the following minimal surface, dicovered by Henneberg, is a stationary solution for a configuration $\langle\Gamma, S\rangle$ satisfying assumption A:

$$x = \cosh(2\lambda u)\cos(2\lambda v) - 1$$
$$y = -\sinh(\lambda u)\sin(\lambda v) - \frac{1}{3}\sinh(3\lambda u)\sin(3\lambda v) \qquad (32)$$
$$z = -\sinh(\lambda u)\cos(\lambda v) + \frac{1}{3}\sinh(3\lambda u)\cos(3\lambda v).$$

It follows from

$$X(u,0) = \{\cosh(2\lambda u) - 1\ ,\ 0, -\sinh(\lambda u) + \frac{1}{3}\sinh(3\lambda u)\}$$
$$= \{2\sin^2(\lambda u)\ ,\ 0\ ,\ + \frac{4}{3}\sinh^3(\lambda u)\}$$

that our minimal surface intersects the plane $y = 0$ in Neil's parabola

$$2x^3 = 9z^2\ ,\qquad y = 0$$

For small values of w we have the expansion

$$x(w) = \mathrm{Re}\ \{2\lambda^2 w^2 + \ldots\}\ ,\ y(w) = \mathrm{Re}\{+2i\lambda^2 w^2 + \ldots\},$$
$$z(w) = \mathrm{Re}\ \{+\frac{4}{3}\lambda^3 w^3 + \ldots\}$$

Therefore, Henneberg's minimal surface is a stationary solution of type I. On account of Theorem 5, it is actually the uniquely determined solution of a minimum problem $P(\Gamma,S)$.

Bibliography

1 R. Courant, Dirichlet's principle, conformal mapping, and minimal surfaces. Interscience, New-York, 1950.

2 J. Frehse, On variational inequalities with lower dimensional obstacles. Bonn 1976, preprint n°. 114, Sonderforschungsbereich 72.

3 J. Frehse, On Signorini's problem and variational problems with thin obstacles. Annali Scuola Norm. Sup. Pisa, Classe di Scienze, Ser. IV, 4, 343-362 (1977).

4 P. Hartman and A. Wintner, On the local behavior of solutions of nonparabolic partial differential equations

Amer. J. Math. 75, 449-476 (1953).

5 S. Hildebrandt and J.C.C. Nitsche, Minimal surfaces with free boundaries. Acta Mathematica 143, 251-272 (1979).

6 S. Hildebrandt and J.C.C. Nitsche, A uniqueness theorem for surfaces of least area with partially free boundaries on obstacles. To appear in : Archive Rat. Mech. Analysis.

7 S. Hildebrandt, Boundary behavior of minimal surfaces. Archive Rat. Mech. Anal. 35, 47-82 (1969).

8 W. Jäger, Behavior of minimal surfaces with free boundaries. Comm. Pure Appl. Math. 23, 803-818 (1970).

9 H. Lewy On minimal surfaces with partially free boundary. Comm. Pure Appl. Math. 4, 1-13 (1951).

10 C.B. Morrey, Multiples integrals in the calculus of variations. Springer, Berlin-Heidelberg - New-York, 1966.

11 J.C.C. Nitsche, Vorlesungen über Minimalflächen.Springer, Berlin Heidelberg - New-York, 1975.

12 J.C.C. Nitsche, The regularity of the trace for minimal surface Annali Scuola Norm. Sup. Pisa Ser. IV, 3, 139-155 (1976).

13 J.E. Taylor, Boundary regularity for various capillarity and boundary problems. Comm. P.D.E. 2, 323-357 (1977).

Stefan HILDEBRANDT

Mathematisches Institut
Der Universität Bonn
Wegelerstrasse 10,
D-5300 BONN

Fed. Rep. of GERMANY

KLAUS KIRCHGÄSSNER & JÜRGEN SCHEURLE

Global branches of periodic solutions of reversible systems

1. INTRODUCTION

A decade ago Crandall and Rabinowitz showed in some remarkable papers [1], [5] that the set of nontrivial solutions of a linear Sturm-Liouville eigenvalue problem is essentially maintained if nonlinear perturbations are added. They proved that in every eigenvalue of the linear problem a branch of solutions bifurcates which exists globally and which preserves the nodal properties well known in the linear case. A few years later, Wolkowisky [9] was able to establish an analogous result for periodic solutions if certain symmetry conditions are satisfied. These symmetry conditions are special forms of a property which we call reversibility, according to G.D. Birckhoff. All these contributions were confined to second order ordinary differential equations, since they relied in one way or another on the use of maximum principles.

Since then, all efforts to extend these results to higher order equations have succeeded under very special circumstances only. In the light of a new uniqueness result, which we proved in [3], we are now able to show global existence of periodic solutions for reversible systems of any order under more general assumptions.

In the course of the investigation, it will become clear for the professional reader, that our method is capable of extensions to partial differential equations. Although we do not intend to make this step explicit in this paper, we present our method, in particular the spaces used, in a way which is suitable for such a generalization. We consider the system of ordinary differential equations

$$(1.1) \qquad \dot{x} - A(\lambda)x = f(\lambda,x)$$

where $x : \mathbb{R} \to \mathbb{R}^n$, $A(\lambda)$ is a real $n \times n$-matrix, depending twice continuously differentiable on the real parameter λ. The function f is supposed to be twice continuously differentiable in λ and x. Moreover we assume that $f(\lambda,0) = 0$, $\nabla_x f(\lambda,0) = 0$ for all λ. System (1.1) is called *reversible* if there exists a reflexion R of $\mathbb{R}^n$ ($R^2 = id$) such that

$$(1.2) \qquad A(\lambda)R = -RA(\lambda), \; f(\lambda,R\circ) = -Rf(\lambda,\circ)$$

holds for all λ. Observe that R has only the semisimple eigenvalues ± 1 and thus may be diagonalized, a fact, which we shall use now and then. In addition, if $x(t)$ is a solution, $Rx(-t)$ is one as well. Observe that (1.2) together with assumption A0 implies that n is even and R has equally many positive as negative eigenvalues. It is an easy exercise to show that then there exists a regular symmetric matrix B which satisfies $BR = -RB$.

Assumption A0 : Denote the spectrum of $A(\lambda)$ by Σ_λ. It is assumed that for some fixed real λ_0

a) re $\mu \neq 0$ for all $\mu \in \Sigma_\lambda$, if $\lambda < \lambda_0$

b) re $\mu = 0$ for exactly two $\mu \in \Sigma_\lambda$, if $\lambda > \lambda_0$.

These eigenvalues are supposed to be simple and are denoted by $\pm i\omega(\lambda)$, where
$\omega: [\lambda_0,\infty) \to [0,\infty)$ is a bijective function which satisfies $\omega'(\lambda) \neq 0$ for all $\lambda \in (\lambda_0,\infty)$.

c) $A(\lambda) = g(\lambda)C(\lambda)$ for $\lambda \geq \lambda_o$ where $g(\lambda) \geq 1$,

$g(\lambda) \to \infty$ as $\lambda \to \infty$,

$$\lim_{\lambda\to\infty} C(\lambda) = C(\infty)$$

where $C(\infty)$ again has exactly two imaginary eigenvalues which are simple.

Observe that for all $|\zeta| > |C(\infty)|$

$$(1.3) \qquad |(C(\infty) - \zeta)^{-1}| \leq \frac{1}{|\zeta|-|C(\infty)|}$$

holds, $\zeta \in \mathbb{C}$, $|C|$ denotes the spectral norm of C corresponding to $|x| = (x\cdot x)^{1/2}$ where $x\cdot y = \sum_k x_k\overline{y_k}$. Assume that for some set $S \subset \mathbb{C}$

$$\sup_{\zeta\in S} |(C(\infty) - \zeta)^{-1}| \leq \Gamma.$$

Then we have for all λ with $|C(\lambda)-C(\infty)| \leq 1/2\Gamma$

$$(1.4) \qquad \sup_{\zeta\in S} |(C(\lambda) - \zeta)^{-1}| \leq 2\Gamma.$$

Given A0, there is a natural decomposition of $A(\lambda)$ into the invariant parts $A_1(\lambda)$, $A_2(\lambda)$

$$A(\lambda) = A_1(\lambda) \oplus A_2(\lambda)$$

where $A_1(\lambda)$ has $\pm i\omega(\lambda)$ as eigenvalues and $A_2(\lambda)$ has the rest of the eigenvalues of $A(\lambda)$ as its spectrum.

Let $\mathbb{H}^\sigma_{loc}(\mathbb{R})$ denote the n-fold product of the real spaces $H^\sigma_{loc}(\mathbb{R})$, where $\sigma \in [0,1]$ and where $H^\sigma_{loc}(\mathbb{R})$ consists of those functions which locally belong to the real Sobolev space $H^\sigma(\mathbb{R})$. For any $p > 0$, we introduce spaces of periodic functions

$$Z_p^\sigma \equiv \{x \in \mathbb{H}_{loc}^\sigma(\mathbb{R}) \;/\; x(t+p) = x(t),$$
$$Rx(-t) = x(t), \text{ a.e.}\}$$

The scalar-products resp. norms of these Hilbert-spaces are standard and will be explicitly defined in section 5. Parallel to the splitting of $A(\lambda)$, there is a decomposition of $Z_p^\sigma = Z_{p,1}^\sigma \oplus Z_{p,2}^\sigma$ via the projections $P(\lambda)$, $Q(\lambda) \equiv id - P(\lambda)$, where $P(\lambda)$ maps into the eigenspace corresponding to $\pm i\omega(\lambda)$. In view of A0, $P(\lambda)$ and $Q(\lambda)$ depend twice continuously differentiable on λ. We set $x_1 \equiv P(\lambda)x$, $x_2 \equiv Q(\lambda)x$.

Assumption A1 :

a) Suppose there exist positive constants γ_o, δ_o, η_o, and $\beta < 1$ such that

$$|f(\lambda,x)| \leq \gamma_o(1+|x|^\beta), \quad x \in \mathbb{R}^n$$

$$|\nabla_x f(\lambda,x)| \leq \delta_o|x| \quad ,|x| < \eta_o$$

holds for all $\lambda \in [\lambda_o,\infty)$.

b) Assume that for each $\lambda \in [\lambda_o,\infty)$

$$-A_2(\lambda)x_2 = f_2(\lambda,x_2)$$

implies $x_2 = 0$.

THEOREM 1.1 (Local version) : Suppose that (1.1) is reversible and that A0 a) and b) hold. Then, for every $\lambda_k = \omega^{-1}(2\pi k/p)$, $(\lambda_k,0)$ is a point of bifurcation in $\mathbb{R} \times Z_p^1$. The nontrivial solutions determine a C^1-branch emanating from $(\lambda_k,0)$.

This theorem is a generalization of the local results in [9] ; however, it is a very special case of a theorem of J. Scheurle in [7], where it is hidden as a trivial case within a class of bifurcating quasiperiodic solutions. In order to keep this paper selfcontained, we shall prove Theorem 1.1 in section 5, nevertheless.

THEOREM 1.2 (Global version) : Let (1.1) be reversible and let A0 as well as A1 be satisfied. Denote by C_p^k, for each $k \in \mathbb{N}$, the component of connectedness in $(\lambda_o,\infty) \times Z_p^1$ which contains the branch of nontrivial solutions emanating from $(\lambda_k,0)$. Then, there exists a positive constant p_o - which may depend on k - such that for all $p \in (0,p_o)$, C_p^k meets the boundary of $(\lambda_o,\infty) \times Z_p^1$.

For second order equations (n = 2) A1 is not needed. Thus, the above theorem contains this case as well, except for the restriction to sufficiently small periods, which stems from the local nature of our uniqueness result. But, for the 2nd-order case its global validity can be proved by elementary methods. Hence, global existence follows then for all positive p.

We even can attach some sort of nodal property to C_p^k by constructing a projection $Z_p^1 \to \mathbb{R}^2$ (see section 5) which gives a one-to-one picture of the solution set near 0. Thus, one is able to define, for each $x \in C_p^k$, a winding number which remains constant (=k) along this component. The proof of the local result is based on the Liapunov-Schmidt method, where the eigenvalue, having originally multiplicity 2, is cut down by means of the reflexion R into an eigenvalue of multiplicity 1. The global result relies on the global bifurcation theorem of Rabinowitz [6] and Theorem 2.1, which we believe to be of interest for its own sake. Therefore, we include a detailed proof in sections 2 to 4 (see also [3] for a different version in a pde-case).

Section 5 finally contains the proofs of the above theorems.

Example : Consider the differential equation

$$(1.5) \qquad y^{(4)} - 2\lambda y'' + (\lambda^2-\lambda^4)y - g(y,y',y'',y''') = 0$$

Assume that g satisfies the following conditions

$$(1.6) \qquad \begin{aligned} &|g(x)| \le \gamma_o(1+|x_1|^{\beta}) \text{ , } x \in \mathbb{R}^4 \text{, some } \beta \in [0,1) \\ &|\nabla g(x)| \le \delta_o|x_1| \qquad , |x_1| < \eta_o \\ &g(x) = (x_1+x_3)h(x) \\ &h(-x_1,x_2,-x_3,x_4) = h(x), \ h(x_1,0,\varepsilon x_1,0) > 0 \end{aligned}$$

for all $x_1 \neq 0$, $\varepsilon > 2$. If (1.6) holds, the conclusions of Theorem 1.2 are valid.

Converting (1.5) into a system $(\lambda^k x_{k+1}=y^{(k)}, \lambda \neq 0)$ of the form (1.1), we obtain

$$C(\lambda) = \begin{pmatrix} 0 & 1 & 0 & 0 \\ 0 & 0 & 1 & 0 \\ 0 & 0 & 0 & 1 \\ 1-\frac{1}{\lambda^2} & 0 & 2/\lambda & 0 \end{pmatrix}, \quad f = \begin{pmatrix} 0 \\ 0 \\ 0 \\ g \end{pmatrix}$$

$$A(\lambda) = \lambda C(\lambda), \ g(\lambda,x) = g(x_1,\lambda x_2,\lambda^2 x_3,\lambda^3 x_4).$$

The system $\dot{x} - A(\lambda)x = f(\lambda,x)$ is reversible with respect to

$$R = \begin{pmatrix} -1 & 0 & 0 & 0 \\ 0 & 1 & 0 & 0 \\ 0 & 0 & -1 & 0 \\ 0 & 0 & 0 & 1 \end{pmatrix}.$$

The eigenvalues of $A(\lambda)$ are

$$\pm\sqrt{\lambda(1+\lambda)} \text{ for } |\lambda| < 1$$

$$\pm i\sqrt{\lambda(\lambda-1)} \text{ and } \pm\sqrt{\lambda(\lambda+1)} \text{ for } |\lambda| > 1.$$

Thus, setting $\lambda_o = 1$, A0 is satisfied as well as A1,a). To show A1,b) we rewrite the above equation in the ξ-system where $x = \xi_1\phi_1 + \ldots + \xi_4\phi_4$, ϕ_j denoting the eigenvectors of $A(\lambda)$, ϕ_1 and ϕ_2 corresponding to $\pm i\sqrt{\lambda(\lambda-1)}$. Setting $\xi_1 = \xi_2 = 0$, we obtain

$$-\sqrt{1+\lambda}\,\xi_3 = \frac{1}{4\sqrt{1+\lambda}}\,g(\lambda,x)$$

(1.7)

$$\sqrt{1+\lambda}\,\xi_4 = -\frac{1}{4\sqrt{1+\lambda}}\,g(\lambda,x)$$

where $\quad x_1 = \xi_3 + \xi_4,\ \sqrt{\lambda}\,x_2 = \sqrt{1+\lambda}\,(\xi_3-\xi_4),$

$$\lambda x_3 = (1+\lambda)(\xi_3+\xi_4),\ \lambda^{3/2}x_4 = (1+\lambda)^{3/2}(\xi_3-\xi_4).$$

Introduce $u = \xi_3 + \xi_4$, $v = \xi_3 - \xi_4$ and conclude easily $v = 0$. Moreover, (1.6) and (1.7) yield

$$-\sqrt{1+\lambda}\,u = \frac{1+\lambda+\lambda^2}{2\sqrt{\lambda+1}}\,u\,h(u,0,(\lambda+\lambda^2)u,0)$$

whence $u = 0$ follows. Therefore A1,b) is fulfilled as well.

We would like to express our gratitude to the people of Waverly Beach for their hospitality, which enabled us to complete this work.

2. LOCAL UNIQUENESS

Let $J(\lambda)$ denote $d/dt - A(\lambda)$ then, in some Y^o-neighbourhood of 0, every solution of (1.1) is uniquely determined by its projection into $\ker J(\lambda)$. This result, which only superficially may seem to be a trivial consequence of a center-manifold theorem (cf. [4]), requires a proof ; since, even if we knew uniqueness of this manifold, solutions of (1.1) in Y^o in general do not belong to this manifold.

The uniqueness theorem will not be stated under the most general assumptions which would imply any finite number of eigenvalues on the imaginary axis. We rather restrict ourselves to the assumptions A0 and A1a) which allow a strengthening of results obtained in [3] towards weaker topologies and uniformity with respect to λ.

The spaces appropriate for our purpose are

$$(2.1) \qquad Y^k \equiv \{x \in \mathbb{H}^k_{loc}(\mathbb{R}) \ / \ E_k(x) < \infty\}, \ k = 0,1$$

where $\mathbb{H}^k$ denotes the n-fold product of the usual Sobolev space H^k and

$$E_k(x) = \sup_{j \in \mathbb{Z}'} \|x\|_{k,j} \qquad , \ \mathbb{Z}' = \mathbb{Z} \setminus \{0\}$$

$$(2.2) \qquad \|x\|_{k,j} \equiv \|x\|_{\mathbb{H}^k(K_j)}$$

$$K_j \equiv \begin{cases} [(j-1)\rho, j\rho] & , \ j \in \mathbb{N} \\ [j\rho, (j+1)\rho] & , \ j \in -\mathbb{N} \end{cases}$$

for some positive ρ. Y^k consists of those vector functions x which are uniformly bounded in $\mathbb{H}^k$.

For $\lambda \geq \lambda_o$ we define the projections

$$P(\lambda) = -\frac{1}{2\pi i}\int_{\Gamma_\lambda} (A(\lambda)-\zeta)^{-1} d\zeta$$

where Γ_λ separates $\pm i\omega(\lambda)$ from the rest of Σ_λ. If Γ_λ is chosen symmetric to the real axis, $P(\lambda)$ acts in real spaces as well. We have $PA = A_1$, $QA = A_2$ where $Q(\lambda) = id - P(\lambda)$. Subsequently it is shown that P and Q are uniformly bounded in λ. Boundedness is clear for compact λ-intervals. Let Γ^o be a positively oriented, smooth Jordan curve separating the imaginary part of the spectrum of $C(\infty)$ from the rest. For all sufficiently large λ we may set $\Gamma_\lambda = g(\lambda)\Gamma^o$ and obtain, using A0,

$$P(\lambda) = -\frac{1}{2\pi i}\int_{\Gamma^o} (C(\lambda) - \zeta_o)^{-1} d\zeta_o$$

whence the boundedness follows by (1.4).

<u>THEOREM 2.1</u> : Let A0 and A1 hold and $\lambda \in [\lambda_o, \infty)$. Then there exists a positive ε, independent of λ such that any two solutions x and y of (1.1) in Y^1 coincide, which satisfy

$$P(\lambda)x(0) = P(\lambda)y(0),\ E_o(x) < \varepsilon,\ E_o(y) < \varepsilon.$$

In particular 0 is E_o-locally unique for $\lambda < \lambda_o$.

This theorem can be proved quite similarly as Theorem 4.4 in [3]. Yet, based on the growth conditions A0,c) and A1,a) we obtain here a stronger result (E_o-small instead of E_1-small and λ-independence).

3. THE LINEAR CASE

In this section we study some properties of the linear part of (1.1). The estimates are used in the final proof of the preceding theorem. The parameter λ always ranges in the interval $[\lambda_o,\infty)$. For the proofs of the subsequent lemmas the following subspaces of $\mathbb{H}^k_{loc}(\mathbb{R})$ are used, consisting of functions having polynomial growth at infinity

$$H^k_r = \{x \in \mathbb{H}^k_{loc}(\mathbb{R}) \ / \ (1+t^2)^{-r/2} D^\alpha x \in H^o(\mathbb{R}),\ 0 \leq \alpha \leq k\}$$

with the norm

$$(3.1) \qquad \|x\|^{k,r} = \sup_{0\leq\alpha\leq k} \{\int_{\mathbb{R}} (1+t^2)^{-r} |D^\alpha x|^2 dt\}^{1/2}$$

The space

$$X^k = \bigcup_{r\in \mathbb{N}_o} H^k_r$$

endowed with the finest locally convex topology such that the injections $H^k_r \hookrightarrow X^k$ are continuous is a bornological and barreled space (cf. [3]). In particular, if a sequence $(x_\ell) \subset X^k$ is bounded and $(x_{\ell|K})$ converges in $H^k(K)$ for every compactum $K \subset \mathbb{R}$ then (x_ℓ) converges in X^k.

Introduce the notation

$$(3.2) \qquad f^\sigma \equiv f(\lambda, \circ+\sigma).$$

<u>LEMMA 3.1</u> : Let be $B(\lambda) = A_2(\lambda) = Q(\lambda)A(\lambda)$ and suppose that A0 holds. Then the mapping $T : f \mapsto x$ from QH^o_r into QH^1_r is well defined for every $r \in \mathbb{N}_o$ by the equation

$$(3.3) \qquad \dot{x} = B(\lambda)x + f$$

and the following estimate is valid

$$\|Tf\|^{0,r} \le \gamma \|f\|^{0,r}$$

where γ is independent of λ (depends on r).

PROOF : Decompose $B = B_+ \oplus B_-$ where B_+ resp. B_- corresponds to the positive resp. negative part of the spectrum. In view of A0 we have $|\mathrm{re}\ \mu| > 2\alpha > 0$ for all μ in the spectrum of $B(\lambda)$, α independent of λ. Hence for compact λ-intervals one obtains by standard arguments

$$\text{(3.4)} \quad \begin{aligned} &\text{a)} \quad |e^{B_+(\lambda)t}| \le c_1 e^{\alpha t} \quad \text{for } t \le 0 \\ &\text{b)} \quad |e^{B_-(\lambda)t}| \le c_1 e^{-\alpha t} \quad \text{for } t \ge 0 \end{aligned}$$

For large λ we argue as in section 2 for $P(\lambda)$. Take Γ_λ to contain the spectrum of $B_+(\lambda)$ in its interior and $\mathrm{re}\ \zeta \ge \alpha$, then

$$e^{B_+(\lambda)t} = -\frac{1}{2\pi i}\int_{\Gamma_\lambda} e^{\zeta t}(B_+(\lambda)-\zeta)^{-1} d\zeta$$

holds. Setting $\Gamma_\lambda = g(\lambda)\Gamma^0$ where Γ^0 contains the positive part of the spectrum of $C_+(\infty)$ we obtain for all sufficiently large λ

$$e^{B_+(\lambda)t} = -\frac{1}{2\pi i}\int_{\Gamma^0} e^{g(\lambda)\zeta_0 t}(C_+(\lambda)-\zeta_0)^{-1} d\zeta_0$$

whence (3.4a) follows via (1.4) and $g(\lambda) \ge 1$ ($\mathrm{re}\zeta_0 \ge \alpha$). Similarly for B_- and (3.4b).

Now the function

$$x(t) = \int_{-\infty}^{t} e^{B_-(\lambda)(t-s)} f_-(s)ds - \int_t^{\infty} e^{B_+(\lambda)(t-s)} f_+(s)ds$$

solves (3.3) ($f = f_+ + f_-$ has polynomial growth). Uniqueness is trivial in view of the spectral assumptions on $B(\lambda)$.

Moreover, this representation for x yields

$$\int_{-\infty}^{\infty}(1+t^2)^{-r}|x(t)|^2 dt \le c_1^2(\|f\|^{o,r})^2 \int_{-\infty}^{\infty}(1+t^2)^{-r}$$

$$\{\int_{-\infty}^{t} e^{-2\alpha(t-s)}(1+s^2)^r ds + \int_t^{\infty} e^{2\alpha(t-s)}(1+s^2)^r ds\} dt$$

whence the inequality for T follows immediately.

LEMMA 3.2 : Let $B(\lambda)$ be as in the preceding Lemma. For fixed $\rho \ge 1$

$$\sup_{k\in\mathbb{Z}'}\|f\|_{o,k} < \infty \text{ implies}$$

$$\sup_{k\in\mathbb{Z}'}\|Tf\|_{o,k} \le \gamma_1 \sup_{k\in\mathbb{Z}'}\|f\|_{o,k}\ , \quad \mathbb{Z}' = \mathbb{Z}\setminus\{0\}$$

where γ_1 is independent of ρ and λ.

PROOF : Using Lemma 3.1, (3.1) and (3.2) we obtain

$$\|Tf^{k\rho}\|_{o,1} \le (1+\rho^2)\|Tf^{k\rho}\|^{o,2} \le (1+\rho^2)\ \gamma\ \|f^{k\rho}\|^{o,2}$$

$$= (1+\rho^2)\ \gamma\ \{\sum_{|j|=1}^{\infty} \|(1+t^2)^{-1} f^{k\rho}\|_{o,j}^2\}^{1/2}$$

$$\le \gamma(1+\rho^2) \sup_{j\in\mathbb{Z}'}\|f^{k\rho}\|_{o,j} \sum_{|j|=1} \frac{1}{(1+(j-1)^2\rho^2)^2}$$

$$\le \gamma_1 \sup_{k\in\mathbb{Z}'}\|f\|_{o,k}\ , \qquad \text{Q.E.D.}$$

LEMMA 3.3 : Let $B(\lambda)$ be as in Lemma 3.1 and $f \in QH_r^o$ with supp $f \subset [\rho,\infty)$. Then the following inequality holds, γ_2 independent of λ and $\rho \ge 1$:

$$\|Tf\|_{o,-1} \leq \gamma_2 e^{-\rho\alpha}\|Tf\|_{o,1}$$

PROOF : For $t < \rho$ and $x = Tf$ one has

$$\dot{x} = B(\lambda)x$$

Since x has polynomial growth at infinity we obtain the representation

$$x(t) = e^{B_+(\lambda)t}x(0)$$

Hence

$$\|x\|_{o,-1}^2 = \int_{-\rho}^{o} |e^{B_+(\lambda)t} x(0)|^2 dt = \int_{o}^{\rho} |e^{B_+(\lambda)(t-\rho)\rho} x(0)|^2 dt$$

$$\leq \gamma_2^2 e^{-2\alpha\rho}\|x\|_{o,1}^2$$

follows. A similar result holds if $\mathrm{supp}\ f \subset [-\infty,-\rho]$.

LEMMA 3.4 : Let be $B(\lambda) = A_1(\lambda) = P(\lambda)A(\lambda)$, $P(\lambda)$ as defined in section 2. Suppose that A0 holds, $f \in PH_r^o$ for some $r \in \mathbb{N}_o$ and $\rho \geq 1$. Then every solution of (3.3) satisfies the inequalities $(\tau = \pm 1)$

(i) $\quad \|x\|_{o,1} \leq \gamma_3\rho\,\|f\|_{o,1} \quad$ if $x(0) = 0$

(ii) $\quad \|x\|_{o,\tau} \leq \gamma_3\rho\,(\|x\|_{o,-\tau} + \|f\|_{o,1} + \|f\|_{o,-1})$

where γ_3 is independent of ρ and λ.

PROOF : It is well known that

$$(B(\lambda)-\zeta)^{-1} = -\frac{P_1(\lambda)}{\zeta-i\omega(\lambda)} - \frac{P_{-1}(\lambda)}{\zeta+i\omega(\lambda)} + H(\zeta)$$

where $P_{\pm 1}$ are the eigenprojections to $\pm i\omega$, H is holomorphic in ζ. Using the

Dunford-Taylor representation of exp $(B(\lambda)t)$ and Cauchy's theorem we obtain

$$e^{B(\lambda)t} = P_1(\lambda)e^{i\omega(\lambda)t} + P_{-1}(\lambda)e^{-i\omega(\lambda)t}$$

Therefore $|\exp(B(\lambda)t)| \leq c_1$ for all t and λ. Inequality (i) follows from

$$x(t) = \int_0^t e^{B(\lambda)(t-s)} f(s)ds$$

by using the Cauchy-Schwarz inequality ; (ii) is obtained for $\tau = 1$ from

$$x(t+\rho) = e^{B(\lambda)\rho}x(t) + \int_t^{t+\rho} e^{B(t+\rho-s)} f(s)ds$$

for $-\rho \leq t \leq 0$. Integration yields

$$\int_{-\rho}^{0} |x(t+\rho)|^2 dt \leq 2c_1^2(\|x\|_{0,-1}^2 + \rho^2\int_{-\rho}^{\rho} |f(s)|^2 ds)$$

and the assertion follows. A similar argument works for $\tau = -1$.

4. THE NONLINEAR CASE

In this section we analyze the nonlinear equation (1.1). Since we have shown all necessary estimates independent of λ and in view of A1a) we may omit the notation of λ in this section.

Choose positive constants a, ρ, δ such that the following inequalities hold

$$\begin{aligned} &6\gamma_1\delta \leq a < 1 \ , \ \delta < 1 \\ (4.1) \qquad &a^2 - a\gamma_2 e^{-\rho\alpha} - 6\gamma_1\delta > 0 \\ &a - \gamma_3\rho(4\delta + a^2) \geq 0, \ \rho \geq 1 \end{aligned}$$

where the γ_j 's are those given in the lemmas of the preceding section, the meaning of ρ is shown in (2.2) and $2\alpha > 0$ is some lower bound for $|\text{re}\ \mu|$ for all eigenvalues μ of A_2. It is easy to see that a and δ can be found if

$\exp(-\rho\alpha) < (\gamma_2\gamma_3\rho^3)^{-1} < 1$ holds.

Setting

$$p(x,y) = P(f(x) - f(y))$$

$$q(x,y) = Q(f(x) - f(y))$$

we conclude from A1a) that $\|x\|_{o,k} < \eta$ and $\|y\|_{o,k} < \eta$ imply ($\eta > 0$ sufficiently small)

$$\|p(x,y)\|_{o,k} \le \delta\|x-y\|_{o,k}$$

(4.2)

$$\|q(x,y)\|_{o,k} \le \delta\|x-y\|_{o,k} \quad , \ k \in \mathbb{Z}' .$$

Observe that $p(x^\sigma,y^\sigma) = p^\sigma(x,y)$ and similarly for q.

Furthermore set

$$(4.3) \qquad \varepsilon_m = a^m\varepsilon_o, \ \varepsilon_o \ge 2\eta$$

and denote

$$f_{|j} = f\chi(K_j)$$

where $\chi(K_j)$ defines the characteristic function of K_j.

4.1. PROPERTY $\Pi(m)$: The pair $x,y \in Y^1$ is said to have property $\Pi(m)$, $m \in \mathbb{N}_o$, if

$$\sup_\sigma \|x^\sigma\|_{o,1} < \eta, \quad \sup_\sigma \|y^\sigma\|_{o,1} < \eta$$

$$\|x-y\|_{o,j} \le \varepsilon_{m+1-|j|}$$

holds for $j = \pm 1, \pm 2, \ldots, \pm(m+1)$.

LEMMA 4.2 : If $x, y \in Y^1$ have property $\Pi(m)$ for some $m \in \mathbb{N}_o$, then the inequalities

$$(4.4) \quad \begin{aligned} &(a) \quad \|T \sum_{j=2\tau}^{\tau\infty} q_{|j}(x,y)\|_{o,\tau} \leq \frac{1}{6}\, \varepsilon_{m+1} \\ &(b) \quad \|T \sum_{j=-\tau}^{\tau\infty} q_{|j}(x,y)\|_{o,\tau} \leq \frac{1}{6}\, \varepsilon_{m+1} \end{aligned}$$

hold for $\tau=+1$ and $\tau=-1$.

PROOF : We proceed by induction. Property $\Pi(0)$ implies $\|q(x,y)\|_{o,j} \leq \varepsilon_o \delta$ for $j \in \mathbb{Z}'$. Using (4.1) and Lemma 3.2, the assertion follows for $m = 0$. Assume, for $\tau=1$, that $\Pi(m-1)$ implies (4.4a). Property $\Pi(m)$ and (4.2) yield

$$\|q(x,y)\|_{o,j} \leq \delta\varepsilon_{m+1-|j|} \;, \quad |j|=1,\ldots,m+1$$

Define

$$\tilde{x} \equiv \sum_{j=3}^{\infty} x_{|j}$$

and $\tilde{y}$ similarly. Observe that $\tilde{x}^{\rho}, \tilde{y}^{\rho}$ possess property $\Pi(m-1)$, whence

$$\|T \sum_{j=2}^{\infty} q_{|j}(\tilde{u}^{\rho}, \tilde{v}^{\rho})\|_{o,1} \leq \frac{1}{6}\, \varepsilon_m$$

Since the argument of T in the above inequality vanishes in K_j, $j \leq 1$, one concludes, using Lemma 3.3

$$\|T \sum_{j=3}^{\infty} q_{|j}(x,y)\|_{o,1} = \|T \sum_{j=2}^{\infty} q_{|j}^{\rho}(\tilde{x},\tilde{y})\|_{o,-1} \leq$$

$$\leq \gamma_2 e^{-\rho\alpha} \; \|T \sum_{j=2}^{\infty} q_{|j}(\tilde{x}^{\rho}, \tilde{y}^{\rho})\|_{o,1} \leq \frac{\gamma_2}{6}\, \varepsilon_m e^{-\rho\alpha}$$

Moreover, $\Pi(m)$ and Lemma 3.2 yield

$$\|Tq_{|2}(x,y)\|_{o,1} \leq \gamma_1 \delta \varepsilon_{m-1}$$

Hence by (4.1)

$$\|T \sum_{j=2}^{\infty} q_{|j}(x,y)\|_{o,1} \leq \gamma_1 \delta \varepsilon_{m-1} + \frac{\gamma_2}{6} \varepsilon_m e^{-\rho\alpha} \leq \frac{1}{6} \varepsilon_{m+1}$$

Similarly, one shows (4.4a) for $\tau = -1$.

For the proof of (4.4b), $\tau = 1$, define $x_-^\rho \equiv x^\rho \chi_-$, $y_-^\rho \equiv y^\rho \chi_-$, where $\chi_- \equiv \chi(\mathbb{R}^-)$, and observe that x_-^ρ, y_-^ρ possess property $\Pi(m)$. Therefore, (4.4a) yields

$$\|T \sum_{j=-1}^{-\infty} q_{|j}(x,y)\|_{o,1} = \|T \sum_{j=-2}^{-\infty} q_{|j}(x_-^\rho, y_-^\rho)\|_{o,-1} \leq \frac{1}{6} \varepsilon_{m+1}$$

Inequality (4.4b) follows similarly.

LEMMA 4.3 : Property $\Pi(m)$ implies

$$\|Tq(x,y)\|_{o,\pm 1} \leq \frac{1}{2} \varepsilon_{m+1}, \quad m \in \mathbb{N}_o.$$

PROOF : Property $\Pi(m)$ and (4.2) yield $\|q_{|1}\|_{o,1} \leq \delta\varepsilon_m$ and by Lemma 3.2 : $\|Tq_{|1}\|_{o,1} \leq \gamma_1 \delta \varepsilon_m$.

Therefore, using Lemma 4.2, one concludes via (4.1)

$$\|Tq\|_{o,1} \leq \|Tq_{|1}\|_{o,1} + \|T \sum_{j=2}^{\infty} q_{|j}\|_{o,1} + \|T \sum_{j=-1}^{-\infty} q_{|j}\|_{o,1}$$

$$\leq \gamma_1 \delta \varepsilon_m + \frac{1}{3} \varepsilon_{m+1} \leq \frac{1}{2} \varepsilon_{m+1}$$

LEMMA 4.4 : Let $m \in \mathbb{N}_o$ be fixed. Any two solutions x,y of (1.1) with $Px(0) = Py(0)$, having property $\Pi(m)$, satisfy

$$\|P(x-y)\|_{o,j} \le \frac{\varepsilon_{m+2-|j|}}{2}, \quad |j| = 1,\dots, m+2.$$

PROOF : We conclude from (1.1)

$$J(\lambda)P(x-y) = p(x,y)$$

whence, via Lemma 3.4,ii) ($\tau = \pm 1$)

$$(4.5) \qquad \|P(x-y)\|_{o,\tau} < \gamma_3 \rho(\delta \|x-y\|_{o,1} + \delta \|x-y\|_{o,-1} + \|P(x-y)\|_{o,-\tau})$$

We show inductively that for τ = sign j

$$(4.6) \qquad \|P(x^{j\rho}-y^{j\rho})\|_{o,\tau} \le \frac{1}{2}\, \varepsilon_{m+1-|j|} \text{ for } |j| = 0,\dots, m+1$$

The case j=0 follows from (4.5), Lemma 3.4,i) and $\gamma_3 \rho\delta\varepsilon_m \le \varepsilon_{m+1}/2$. Assume that (4.6) holds for some $|j-\tau| \le m$. Take $j > 0$. Since $x^{j\rho}$, $y^{j\rho}$ have property $\Pi(m-j)$, one obtains

$$\|P(x^{j\rho}-y^{j\rho})\|_{o,-1} = \|P(x^{(j-1)\rho}-y^{(j-1)\rho})\|_{o,1}$$

$$\le \frac{1}{2}\, \varepsilon_{m+1-(j-1)}$$

Now, (4.1), (4.5) for $\tau = 1$ and (4.6) yield

$$\|P(x-y)\|_{o,j+1} = \|P(x^{j\rho}-y^{j\rho})\|_{o,1} \le$$

$$\le \gamma_3 \rho(2\delta\varepsilon_{m-j} + \frac{1}{2}\, \varepsilon_{m+2-j}) \le \frac{1}{2}\, \varepsilon_{m+1-j}.$$

A similar argument holds for negative τ and j.

Thus, the lemma is proved.

PROOF OF THEOREM 2.1 : We show inductively that x,y have property $\Pi(m)$ for all $m \in \mathbb{N}_o$, yielding $\|x-y\|_{o,j} = 0$ for all integers $j \neq 0$, and thus $x = y$.

With $\varepsilon = \eta \leq \varepsilon_o/2$, $\pi(0)$ follows. Assume, x,y have $\Pi(m-1)$; then $x^{j\rho}$, $y^{j\rho}$ have $\Pi(m-1-|j|)$ for $|j| = 0,1,\ldots, m-1$. Since x^j , y^j solve (1.1)

$$Q(x^{j\rho}-y^{j\rho}) = Tq(x^{j\rho},y^{j\rho})$$

holds whence, by Lemma 4.3, one obtains

$$\|Q(x-y)\|_{o,j+1} = \|Q(x^{j\rho}-y^{j\rho})\|_{o,1} \leq \frac{1}{2}\,\varepsilon_{m-|j|} \text{ for } 0 \leq j \leq m$$

$$\|Q(x-y)\|_{o,j-1} = \|Q(x^{j\rho}-y^{j\rho})\|_{o,-1} \leq \frac{1}{2}\,\varepsilon_{m-|j|} \text{ for } 0 \geq j \geq -m$$

If $\ker J(\lambda) \neq \{0\}$, lemma 4.4 implies

$$\|P(x-y)\|_{o,j+\tau} \leq \frac{1}{2}\,\varepsilon_{m-|j|}\ ,\quad |j| = 0,\ldots, m$$

If $\ker J(\lambda)$ is trivial, Q represents the identity. In both cases we have

$$\|x-y\|_{o,j} \leq \varepsilon_{m+1-|j|}\ ,\quad |j| = 1,\ldots, m+1$$

Therefore x,y have property $\Pi(m)$ and the theorem is proved.

5. GLOBAL EXISTENCE OF PERIODIC SOLUTIONS

In this section we prove the Theorems 1.1 and 1.2. As simple examples show, reversibility is needed for the existence of periodic solutions. Also A0 is a quite natural assumption, whereas A1 is artificially chosen such that Theorem 2.1 can be applied.

Let Z_p^σ be the space defined in section 1, $\sigma \in [0,1]$, with the scalar products and norms

$$(x,y)_p = \frac{1}{p}\int_0^p x(t)\cdot\overline{y(t)}dt, \quad |x|_p = (x,x)_p^{1/2}$$

$$((x,y))_p = (x,y)_p + (\dot{x},\dot{y})_p, \quad \|x\|_p = ((x,y))_p^{1/2}$$

The local case

For the proof of Theorem 1.1 we simply apply a result of Crandall ([O], Theorem 1). Observe that Z_p^σ, $\sigma > 1/2$, is continuously imbedded in $C_0^0(\mathbb{R})$, the space of bounded continuous functions equipped with the topology of uniform convergence. Since $f \in C^2(\mathbb{R}^n)$, f maps Z_p^σ into Z_p^0 and is twice continuously differentiable ($\sigma > 1/2$ is assumed throughout).

We solve

$$J(\lambda)x - f(\lambda,x) = 0, \quad J(\lambda) \equiv \frac{d}{dt} - A(\lambda). \tag{5.1}$$

For $\omega(\lambda_k) = 2\pi k/p$, $k \in \mathbb{N}$, $J(\lambda_k)$ has a 1-dimensional kernel spanned by

$$x_0 = \phi_1(\lambda_k)e^{i\omega(\lambda_k)t} + \phi_{-1}(\lambda_k)e^{-i\omega(\lambda_k)t}$$

here $\phi_{\pm 1}(\lambda)$ are normalized eigenfunctions of $A(\lambda)$ to $\pm i\omega(\lambda)$. According to [O] one has to show that

$$J(\lambda_k)z + \tau J'(\lambda_k)x_o$$

is an isomorphism from $\mathbb{R} \times Z$ on Z_p^o, where Z is a suitable, closed subspace of Z_p^1. Choose for Z the orthogonal complement of x_o and set

$$y_o = i(g^1(\lambda_k)e^{i\omega(\lambda_k)t} - g^{-1}(\lambda_k)e^{-i\omega(\lambda_k)t})$$

where $g^{\pm 1}(\lambda)$ are the adjoint eigenfunctions satisfying $\phi_j \cdot g^k = \delta_j^k$. Then the range of $J(\lambda_k)$ is the orthogonal complement of y_o, and since

$$(J'(\lambda_k)x_o, y_o)_p = - (A'(\lambda_k)x_o, y_o)_p = - 2\omega'(\lambda_k) \neq 0$$

according to A0, $J'(\lambda_k)x_o \notin$ range $J(\lambda_k)$, what had to be proved. Hence, each $(\lambda_k, 0)$ is a bifurcation point in $\mathbb{R} \times Z_p^1$ and in some neighbourhood of this point the non-trivial solutions form a C^1-curve

$$x_k(\varepsilon) = \varepsilon(x_o + z(\varepsilon))$$

(5.2)

$$\lambda_k(\varepsilon) = \lambda_k + \tau(\varepsilon)$$

with $z(0) = 0$, $\tau(0) = 0$.

The global case

To prove Theorem 1.2 take any real nonsingular $n \times n$-matrix B with $BR = -RB$ (exists since n must be even and trace $R = 0$) and define

$$L = \frac{d}{dt} - B$$

L is a topological isomorphism between Z_p^1 and Z_p^o.

Hence, (5.1) is equivalent to

$$(5.3) \qquad x = M(\lambda)x - L^{-1}f(\lambda,x), \; M(\lambda) = L^{-1}(A(\lambda)-B)$$

with $x \in Z_p^1$. $M(\lambda)$ has at $\lambda = \lambda_k$ the geometrically simple eigenvalue 1 with x_o being an eigenvector. The range is the orthogonal complement of L^*y_o (L^* = adjoint of L). Define $\rho(\lambda)$ to be the eigenvalue of $M(\lambda)$ with $\rho(\lambda_k) = 1$; then

$$\rho'(\lambda_k)(Lx_o,y_o)_p = (A'(\lambda_k)x_o,y_o)_p = 2\omega'(\lambda_k)$$

whence $\rho'(\lambda_k) \neq 0$ via A0. Hence the global result of Rabinowitz [6] implies that the component C_p^k of nontrivial solutions bifurcating at $(\lambda_k,0)$ either "meets" the boundary of $(\lambda_o,\infty) \times Z_p^\sigma$ or another bifurcation point (in Z_p^σ all operators are completely continuous). Observe that C_p^k is closed and connected in $(\lambda_o,\infty) \times Z_p^1$ (use (5.3) and the compact imbedding of Z_p^1 into Z_p^σ) and obeys the same alternatives in this smaller space.

We denote the λ-dependence of any solution of (5.1) explicitly by $x(\circ,\lambda)$ and define the projection $\tilde{P}(\lambda)$ coordinatewise via

$$\tilde{P}(\lambda)x(\circ,\lambda) = x(\circ,\lambda)\cdot y^1(\lambda)$$

which maps the $\mathbb{R}^n$-valued functions in Z_p^1 into $\mathbb{C}$-valued H_{loc}^1-functions. Since $x.g^1 = \overline{x.g^{-1}}$ we have in particular that $\tilde{P}x = 0$ if and only if $Px = 0$. For the following lemma take ε from Theorem 2.1.

<u>LEMMA 5.1</u> : Let $x(\circ,\lambda) \in Z_p^1$ solve (5.1) for some λ and suppose that

$$E_o(x) < \varepsilon, \; \tilde{P}(\lambda)x(t_o,\lambda) = 0$$

holds for some $t_o \in \mathbb{R}$; then $x = 0$.

PROOF : Set $y(t) = x(t+t_o,\lambda)$, then y solves (1.1), belongs to H^1_{loc} and is p-periodic. Moreover we have $E_o(y) < \varepsilon$ and $P(\lambda)y(0) = 0$. Hence Theorem 2.1 implies $y = 0$ and thus $x = 0$.

LEMMA 5.2 : Let $x(\circ,\lambda) \in Z^1_p$ be a solution of (5.1) which satisfies

$$\tilde{P}(\lambda)x(t_o,\lambda) = 0$$

for some $t_o \in \mathbb{R}$. Then the following inequalities hold for $x_1 = P(\lambda)x(\circ,\lambda)$, $x_2 = Q(\lambda)x(\circ,\lambda)$

$$\text{a)} \quad |x_1|_p^2 \le \gamma_4 p^2 + \gamma_5 p^{2-\beta}(|x_1|_p^{2\beta} + |x_2|_p^{2\beta})$$

(5.4)

$$\text{b)} \quad |x_2|_p^2 \le \gamma_6 + \gamma_7(|x_1|_p^{2\beta} + |x_2|_p^{2\beta})$$

where all γ_j are independent of λ.

PROOF : Since x_1 satisfies

$$x_1(t) = \int_{t_o}^{t} e^{A_1(\lambda)(t-s)} f_1(\lambda,x_1+x_2)(s)ds$$

and since $|\exp(A_1(\lambda)t)| \le c_1$ for all λ and t, the first inequality follows by using Hölder's inequality for $p = 2/\beta$. To obtain b) we invert $J_2(\lambda)x_2 = f_2$. Denoting by x_ν resp. f_ν the νth. Fourier-coefficient of x resp. f we have

$$x_{2\nu} = R_\nu(\lambda)f_{2\nu}$$

where

$$R_\nu(\lambda) = (\frac{2\pi i\nu}{p} - A_2(\lambda))^{-1} .$$

Hence it remains to be shown that $R_\nu(\lambda)$ is uniformly bounded in ν and λ. Since $A_2(\lambda)$ is nonsingular we may write

$$R_\nu^+(\lambda) = -\frac{1}{2\pi i}\int_{\Gamma_\lambda} \frac{1}{\zeta}\left(\frac{2\pi i\nu}{p} - A_2^+(\lambda) - \zeta\right)^{-1} d\zeta$$

and a similar expression for $A_2^-(\lambda)$; Γ_λ surrounds the positive resp. negative part of the spectrum of $A_2(\lambda)$. It is not hard to show that for every $\lambda \in [\lambda_o,\infty)$ Γ_λ can be chosen to be a line parallel to the imaginary axis with re $\zeta = +\alpha$ resp. $-\alpha$ $(\alpha > 0)$. Hence we have

$$R_\nu^+(\lambda) = -\frac{1}{2\pi}\int_{-\infty}^{\infty} \frac{1}{\alpha - i\tau}\left(\frac{2\pi i\nu}{p} - A_2^+(\lambda) - \alpha + i\tau\right)^{-1} d\tau.$$

Setting $s = \tau + 2\pi\nu/p$ one sees that $\nu R_\nu^+(\lambda)$ is bounded in ν in compact λ-intervals. To obtain an estimate for large λ we substitute $\tau = g(\lambda)s - 2\pi\nu/p$ and obtain

$$R_\nu^+(\lambda) = \frac{1}{2\pi}\int_{-\infty}^{\infty} \frac{1}{\alpha - i(g(\lambda)s - 2\pi\nu/p)}\left(C_2^+(\lambda) + \frac{\alpha}{g(\lambda)} - is\right)^{-1} ds$$

whence the boundedness follows from A0 and (1.4).
A similar consideration for $R_\nu^-(\lambda)$ completes the proof of uniform boundedness of $R_\nu(\lambda)$ in ν and λ. Now, (5.4b) follows via Parseval's equality using Hölder's inequality for $p = 2/\beta$.

For any $\rho \geq 1$ we define - as in section 2 - $K_1 = [0,\rho]$. If $x \in Z_p^o$ we obtain easily for every $p < \rho$

$$(5.5) \qquad \sqrt{\rho - p}\ |x|_p \leq E_o(x) \leq \sqrt{\rho + p}\ |x|_p$$

<u>LEMMA 5.3</u> : Let p_n be a sequence of positive numbers converging to 0 ; suppose $x_n \in Z_{p_n}^1$ are solutions of (5.1) resp. (1.1) for $\lambda = \lambda_n$. Assume that for $t_n \in \mathbb{R}$

$$\tilde{P}(\lambda_n)x_n(t_n,\lambda_n) = 0.$$

Then $E_o(x_n)$ converges to 0.

PROOF : Apply Lemma 5.2. Inequality (5.4a) yields that $|x_{1n}|_{p_n}$ is bounded by $|x_{2n}|_{p_n}^{\beta}$, whence it follows from (5.4b) that $|x_{2n}|_{p_n}$ is bounded. Using (5.4a) again we conclude that $|x_{1n}|_{p_n} \to 0$.

Set $g = g(\lambda)$, $s = gt$, $y(s) = x(gt)$. Elementary calculations yield

$$(5.6) \qquad |x|_p = |y|_{pg}, \quad |\dot{x}|_p = g|y'|_{pg}$$

Moreover y satisfies

$$(5.7) \qquad y' - C(\lambda)y = \frac{1}{g(\lambda)} f(\lambda, y)$$

where $C(\lambda)$ was defined in A0. We can use the integral representation of y which was already applied in the proof of Lemma 3.1

$$g\, y_2(s) = \int_{-\infty}^{s} e^{C_2^-(\lambda)(s-\sigma)} f_-(\sigma)ds - \int_{s}^{\infty} e^{C_2^+(\lambda)(s-\sigma)} f_+(\sigma)ds$$

where $C_2 = QC$ and C_2^+, C_2^- denote the parts of C_2 having positive resp. negative spectrum. Since for all eigenvalues μ of C_2 we have $|\text{re}\,\mu| > 2\alpha > 0$ for some positive α and all λ we conclude

$$g\, |y_2|_{pg} \le c_1 |f|_{pg}.$$

In view of A1,a) and (5.6), $|f|_{pg}$ is uniformly bounded for all y_n, λ_n ; hence $g_n |y_{2n}|_{p_n g_n}$ is bounded as well ($g_n = g(\lambda_n)$). But (5.7) implies together with A0,c) that

$$g_n |y'_{2n}|_{p_n g_n} = |\dot{x}_{2n}|_{p_n}$$

is bounded. Quite similarly as in (5.5) one shows that (see (2.2))

$$E_1(x) \le \sqrt{\rho + p}\,||x||_p .$$

Hence $E_1(x_{2n})$ is bounded. Therefore, we may select a subsequence convergent in $\mathbb{H}^\sigma(0,\rho)$ for $1 > \sigma > 1/2$ to some x_2. Since this convergence is uniform in $[0,\rho]$ and, in view of the periodicity, also in $\mathbb{R}$. Moreover, x_{2n} being p_n-periodic, x_2 has to be a constant. It follows from

$$x_{1n}(t) = \int_{t_n}^{t} e^{A_1(\lambda_n)(t-s)} f_1(\lambda_n, x_{1n}+x_{2n})(s)ds$$

by using Hölder's inequality and A1,a)

$$|x_{1n}(t)| \le c_1 \sqrt{p_n}\,(1+p_n^{\frac{1-\beta}{2}}\,(|x_{1n}|_{p_n}^{\beta} + |x_{2n}|_{p_n}^{\beta}))$$

and therefore $x_{1n} \to 0$ uniformly on $\mathbb{R}$.

If λ_n contains a convergent subsequence we obtain $(\lambda_n \to \lambda)$

$$- A_2(\lambda)x_2 = f_2(\lambda, x_2)$$

and hence $x_2 = 0$ by A1,b). Otherwise $\lambda_n \to \infty$; then $g(\lambda_n) \to \infty$ by A0,c), and it follows $C_2(\infty)x_2 = 0$. Thus we again obtain $x_2 = 0$. Hence we have $E_o(x_n) \to 0$ for every convergent subsequence which proves the theorem.

<u>PROOF OF THEOREM</u> 1.2 : For each $(\lambda,x) \in Z_p^1$

$$\tilde{P}(\lambda)x(t,\lambda) = z(t),\ t \in [0,\frac{2\pi}{p}]$$

defines a closed curve $\Gamma_{x,\lambda}$ in C. Its winding number with respect to 0 is defined by

$$w_{x,\lambda} = \frac{1}{2\pi i}\int_{\Gamma_{x,\lambda}} \frac{dz}{z}$$

if $0 \notin \Gamma_{x,\lambda}$. According to Theorem 1.1 there exists a neighbourhood of $(\lambda_k, 0)$ in $\mathbb{R} \times Z^1_p$ such that all nontrivial solutions have the form (5.2). Therefore $w_{x,\lambda}$ is well defined in this neighbourhood and equals k. Hence if C^k_p, meets another bifurcation point there must exist $(\lambda,x) \in C^p_k$, $t_o \in \mathbb{R}$ such that $x \neq 0$,

$$\tilde{P}(\lambda)x(t_o,\lambda) = 0.$$

Assume, Theorem 1.2 were false. Then there exist sequences $p_n \to 0$, $t_n \in \mathbb{R}$ $(\lambda_n, x_n) \in C^k_{p_n}$, $x_n \neq 0$, $\lambda_n \in (\lambda_o, \infty)$ such that

$$\tilde{P}(\lambda_n)x_n(t_n,\lambda_n) = 0$$

According to Lemma 5.3 we have $E_o(x_n) \to 0$. Thus, for some n_o and all $n \geq n_o$, $E_o(x_n) < \varepsilon$ and $P(\lambda_n)x_n(t_n) = 0$. Hence, by Lemma 5.1, $x_n = 0$ which leads to a contradiction.

Bibliography

0 M.G. Crandall, An introduction to constructive aspects of bifurcation and the implicit function theorem in: Applications of bifurcation theory (P.H. Rabinowitz ed.), Academic Press (1977), 1-36.

1 M.G. Crandall - P.H. Rabinowitz, Nonlinear Sturm-Liouville eigenvalue problems and topological degree, J. Math. Mech. 19 (1969/70), 1083-1102.

2 C. Foias - G. Prodi, Sur le comportement global des solutions non-stationnaires des équations de Navier-Stokes en dimension 2, Rend. Sem. Mat. Univ. Padova 39 (1967), 1-34.

3 K. Kirchgässner - J. Scheurle, On the bounded solutions of a semilinear equation in a strip. J. Diff. Equ. 32 (1979), 119-148.

4 J.E. Marsden - M. McCracken, The Hopf Bifurcation and its Application, Springer Verlag, Berlin - Heidelberg - New York (1976).

5 P.H. Rabinowitz, Nonlinear Sturm-Liouville problems for second order ordinary differential equations, Comm. Pure Appl. Math. 23 (1970), 939-961.

6 P.H. Rabinowitz, Some global results for nonlinear eigenvalue problems, J. Funct. Anal. 7 (1971), 487-513.

7 J. Scheurle, Verzweigung quasiperiodischer Lösungen bei reversiblen dynamischen Systemen, Habilitationsschrift, Stuttgart (1980)

8 F. Trèves, Topological vector spaces, distributions and kernels, Academic Press, New York (1967).

9 J.H. Wolkowisky, Branches of periodic solutions of the nonlinear Hill's equation, J. Diff. Equ. 11 (1972), 385-400.

Klaus KIRCHGÄSSNER and Jürgen SCHEURLE

Math. Institut A, Universität Stuttgart
Pfaffenwaldring 57,
D-7000 STUTTGART 80
Federal Republic of GERMANY

JEAN MAWHIN

Perturbations non-linéaires d'opérateurs linéaires à noyau de dimension infinie

1. INTRODUCTION

Ce travail, qui rend compte de résultats obtenus en collaboration avec M. Willem [6] , est consacré à l'étude de la résolubilité d'équations de la forme

$$Lx - Nx = f \tag{1.1.}$$

dans un espace de Hilbert réel H, où $f \in H$, $L : \text{dom}\, L \subset H \to H$ est linéaire et $N : H \to H$ est en général non-linéaire. On s'intéresse au cas dit "résonnant", c'est-à-dire tel que

$$\text{Ker } L \neq \{0\} .$$

Dans la ligne d'un récent travail de Brézis-Nirenberg [2] , et contrairement à la plupart des résultats obtenus jusqu'à présent, on s'interessera particulièrement au cas où

$$\dim \text{Ker } L = \infty. \tag{1.2.}$$

Un tel cas se présente en particulier dans la formulation abstraite du problème des solutions périodiques d'équations d'ondes non-linéaires ou de certains problèmes aux limites pour des équations différentielles dans des espaces de dimension infinie.

On démontrera, pour une classe relativement vaste d'équations de type (1.1.), un théorème de continuation du type de Leray-Schauder dont on pourra

déduire assez aisèment le résultat de Brézis-Nirenberg et une extension au cas où (1.2.) est satisfaite d'un théorème de Cesari-Kannan [3] .
Ce théorème de continuation fait appel à des résultats préliminaires sur les équations de Hammerstein, qui peuvent être intéressants en soi et généralisent un théorème de de Figueiredo-Gupta [4] . Pour des détails complémentaires, nous renverrons le lecteur à [5,6] .

2. EXISTENCE, UNICITE ET DEPENDANCE CONTINUE DES SOLUTIONS D'EQUATIONS DE HAMMERSTEIN

Soit H un espace de Hilbert réel de produit scalaire $(,)$ et de norme $|.|$. Pour simplifier le langage, introduisons la

DEFINITION 2.1. : Une paire (B,F) d'applications de H dans H sera dite h-*comptatible de constantes* a *et* b si les conditions suivantes sont satisfaites

(i) $0 \leq b < a$

(ii) $B(0) = 0$ et $(Bu-Bv,u-v) \geq a|Bu-Bv|^2$ pour tout $u,v \in H$

(iii) F est demi-continu et $(Fu-Fv,u-v) \geq -b|u-v|^2$ pour tout $u,v \in H$.

Le résultat suivant, qui se démontre facilement à partir de la théorie des opérateurs maximaux-monotones, généralise un théorème de de Figueiredo-Gupta [4] .

PROPOSITION 2.1. : Soit (B,F) une paire d'applications h-compatibles de constantes a et b. Alors, pour tout $f \in H$, l'équation de Hammerstein

$$x + BFx = f$$

possède une solution unique x. En outre, on a

$$|x-f| \leq (a-b)^{-1} \; |F(f)|$$

Le résultat de dépendance continue suivant est modelé sur un théorème de Brézis-Browder [1] relatif à une autre classe d'équations de Hammerstein.

PROPOSITION 2.2. : Soit (B,F) et (B_n,F_n), $n \in \mathbb{N}^*$ des couples d'applications de H dans H h-compatibles de constantes a et b et $(f_n)_{n \in \mathbb{N}^*}$ une suite dans H qui converge vers f. Supposons satisfaites les conditions suivantes :

$$\text{pour tout borné } S \subset H, \quad \bigcup_{n \in \mathbb{N}^*} F_n(S) \text{ est borné;} \qquad (1)$$

$$\text{pour tout } u \in H, \; B_n Fu \to BFu \text{ si } n \to \infty ; \qquad (2)$$

$$\text{pour tout } u \in H, \; F_n u \to Fu \text{ si } n \to \infty . \qquad (3)$$

Alors, si

$$u = (I+BF)^{-1} f, \quad u_n = (I+B_n F_n)^{-1} f_n , \quad n \in \mathbb{N}^* ,$$

on a

$$u_n \to u \quad \text{pour } n \to \infty .$$

3. UN THEOREME DE CONTINUATION DU TYPE DE LERAY-SCHAUDER

Soit H un espace de Hilbert réel de produit scalaire $(,)$ et de norme $|.|$, $L : \text{dom } L \subset H \to H$ un opérateur linéaire, fermé, de domaine dense et d'image (fermée) telle que

$$\text{Im } L = (\ker L)^{\perp} .$$

En conséquence $L_{|\text{dom } L \cap \text{Im } L}$ est injective et a pour image $\text{Im } L$, ce qui implique l'existence d'un inverse continu

$$K : \text{Im } L \to \text{dom } L \cap \text{Im } L.$$

On désignera par $P : H \to H$ le projecteur orthogonal sur $\text{Ker } L$.

Soit $N : H \to H$ une application demi-continue, qui transforme les bornés en bornés et telle que :

(i) $I-P - \varepsilon N : H \to H$ est monotone avec $\varepsilon = \pm 1$.

(ii) $K(I-P)N : H \to H$ est compact sur tout borné de H.

On a alors le résultat suivant :

THEOREME 3.1. : Soit L et N vérifiant les conditions ci-dessus. S'il existe un ouvert borné $\Omega \subset H$ tel que $0 \in \Omega$ et

$$Lx + \varepsilon(1-\lambda)Px - \lambda Nx \neq 0 \qquad (3.1.)$$

pour tout $(x,\lambda) \in (\text{dom } L \cap \text{fr } \Omega) \times]0,1[$, alors l'équation

$$Lx - Nx = 0 \qquad (3.2.)$$

possède au moins une solution.

DEMONSTRATION : Elle se fait en deux étapes.

1. *Pour tout* $\lambda \in]0,1]$, *l'équation*

$$Lx + \varepsilon(1-\lambda)Px - \lambda Nx = 0 \qquad (3.3.)$$

possède au moins une solution dans dom $L \cap \Omega$.

Par un résultat connu, l'équation (3.3.) est équivalente à l'équation

$$x - Px = [\varepsilon P + K(I-P)] (-\varepsilon(1-)Px + \lambda Nx)$$

c'est-à-dire à l'équation

$$x - \lambda P(P+\varepsilon N)x = \lambda K(I-P)Nx. \qquad (3.4.)$$

Considérons, pour $\mu \in [0,1]$, la famille d'équations

$$x - \mu\lambda P(P+\varepsilon N)x = \mu\lambda K(I-P)Nx. \qquad (3.5.)$$

On vérifie sans peine que pour chaque $\mu \in [0,1]$, le couple $(\lambda P, -\mu(P+\varepsilon N))$ est h-compatible de constantes λ^{-1} et 1. Dès lors, en posant

$$T_{\mu}^{\lambda}(x) = I - \mu\lambda P(P+\varepsilon N),$$

il résulte de la Proposition 2.1. que (3.5.) est équivalent à

$$x = (T_{\mu}^{\lambda})^{-1} [\mu\lambda K(I-P)Nx] = R_{\lambda}(x,\mu).$$

Par ailleurs, trivialement,

$$R_{\lambda}(.,0) = 0$$

et, grâce aux hypothèses et à la Proposition 2.2., on peut vérifier que R_{λ} est compacte sur tout borné de $H \times [0,1]$. Comme, enfin, le fait que $0 \in \Omega$,

(3.1.) et la chaîne d'équivalences ci-dessus impliquent que

$$x - R_\lambda(x,\mu) \neq 0$$

pour tout $(x,\mu) \in \mathrm{fr}\Omega \times [0,1]$, la thèse résulte du théorème de Leray-Schauder.

2. *L'équation* (3.2.) *possède au moins une solution.*

Soit $(\lambda_n)_{n \in \mathbb{N}^*}$ une suite dans $]0,1[$ telle que $\lambda_n \to 1$ si $n \to \infty$. Par la première partie de la démonstration, il existe $(x_n)_{n \in \mathbb{N}^*}$ dans $\mathrm{dom}\, L \cap \Omega$ telle que

$$Lx_n + \varepsilon(1-\lambda_n)Px_n - \lambda_n Nx_n = 0 , \quad n \in \mathbb{N}^* \qquad (3.6.)$$

et dès lors

$$y_n = \lambda_n K(I-P)Nx_n$$

$$\varepsilon(1-\lambda_n)z_n = \lambda_n PNx_n , \quad n \in \mathbb{N}^* .$$

en posant

$$(I-P)x_n = y_n , \quad Px_n = z_n , \quad n \in \mathbb{N} .$$

Comme (z_n) et (Ly_n) sont bornées et que (y_n) est contenue dans un compact, on peut supposer, à une extraction de sous-suite près, que, pour $n \to \infty$,

$$y_n \to y , \quad z_n \rightharpoonup z \quad \text{et} \quad Ly_n \rightharpoonup Ly .$$

Par la monotonie de $I-P-\varepsilon N$, on a, pour tout $v \in H$, et en utilisant (3.6.),

$$(y_n - \lambda_n^{-1}(1-\lambda_n)z_n - \lambda_n^{-1}\varepsilon Ly_n - (I-P-\varepsilon N)v, x_n - v) \geq 0$$

et donc, par passage à la limite

$$(y-\varepsilon Ly - (I-P-\varepsilon N)v, x-v) \geq 0 ,$$

pour tout $v \in H$. $I-P-\varepsilon N$ étant maximal monotone, cela implique

$$y - \varepsilon Ly = (I-P-\varepsilon N)x = y - \varepsilon Nx$$

et dès lors la thèse.

4. APPLICATIONS A DES RESULTATS ABSTRAITS DU TYPE LANDESMAN-LAZER

Donnons tout d'abord une extension au cas ou $\dim \operatorname{Ker} L$ est non finie d'un théorème de Cesari-Karman [3] .

THEOREME 3.2. : Supposons que L et N vérifient les conditions énoncées au début du paragraphe 3. Si en outre :

(i) il existe $r > 0$ tel que $|(K(I-P)Nx| \leq r$ pour tout $x \in H$;

(ii) il existe $R > 0$ tel que $\varepsilon(Nx,Px) \leq 0$ pour tous les x tels que $|Px| = R$ et $|(I-P)x| \leq r$, alors l'équation (3.2.) possède au moins une solution.

DEMONSTRATION : Soit

$$\Omega = \{x \in H : |Px| < R \quad \text{et} \quad |(I-P)x| < r\}$$

de telle sorte que $0 \in \Omega$ et

$$\operatorname{fr} \Omega = S_1 \cup S_2$$

où

$$S_1 = \{x \in H : |Px| = R \quad \text{et} \quad |(I-P)x| \leq r\}$$

$$S_2 = \{x \in H : |Px| \leq R \quad \text{et} \quad |(I-P)x| = r\}.$$

Pour chaque $\lambda \in]0,1[$, l'équation (3.4.) , et donc (3.3.), équivaut à

$$x - Px = \lambda K(I-P)Nx \qquad (3.7.)$$

$$-(1-\lambda)Px + \varepsilon\lambda PNx = 0 \qquad (3.8.)$$

et dès lors, pour toute solution éventuelle de (3.3.), on a, par (3.7.),

$$|(I-P)x| \leq \lambda r < r ,$$

c'est-à-dire $x \notin S_2$. D'ailleurs, (3.8.) implique

$$-(1-\lambda)\ |Px|^2 + \varepsilon\lambda(Nx,Px) = 0$$

ce qui est impossible, par (ii), pour $x \in S_1$. Donc les solutions éventuelles de (3.3.) sont telles que $x \notin \text{dom } L \cap \text{fr } \Omega$ et la thèse résulte du Théorème 3.1.

Appliquons maintenant ce Théorème 3.1. à la démonstration d'un résultat de Brézis-Nirenberg [2] . Supposons que L vérifie les hypothèses du début du paragraphe 3. Cela implique aisèment que, pour tout $x \in \text{dom } L$,

$$(Lx,x) \geq -|K|\ |Lx|^2 \qquad (3.9.)$$

et on désignera par θ la plus grande constante positive telle que

$$(Lx,x) \geq -\theta^{-1}\ |Lx|^2$$

pour tout $x \in \text{dom } L$. Soit $B : H \to H$ une application demi-continue qui transforme les bornés en bornés et telle que

(i) $I-P+B$ est monotone;

(ii) $K(I-P)B$ est compacte sur tout borné de H.

On a alors le

THEOREME 3.3. : Dans les conditions ci-dessus, s'il existe $\gamma \in]0,\theta[$ tel que, pour tout $x \in H$ et tout $y \in H$,

$$(Bx-By,x) \geq \gamma^{-1}|Bx|^2-c(y) \tag{3.10.}$$

où $c(y)$ dépend seulement de y, alors

$$\text{int}(\text{Im } L+\text{conv Im } B) \subset \text{Im}(L+B).$$

En particulier, si B est surjective, $L+B$ est surjective.

DEMONSTRATION : Soit

$$f \in \text{Int}(\text{Im } L+\text{conv Im } B)$$

et $N = f-B$. En vertu des hypothèses faites et du Théorème 3.1., il suffit de montrer que l'ensemble des solutions de la famille d'équations

$$Lx + (1-\lambda)Px + \lambda Bx = \lambda f \tag{3.11.}$$

est borné a priori indépendamment de $\lambda \in]0,1[$. Pour tout $h \in H$ de norme suffisamment petite, on peut écrire

$$f+h = Lv + \sum_{i=1}^{n} t_i Bw_i$$

où $v \in \text{dom } L$, $w_i \in H$, $t_i \geq 0$ $(1 \leq i \leq n)$ et $\sum_{i=1}^{n} t_i = 1$.

Dès lors, (3.10.) devient

$$(1-\lambda)Px + \lambda(Bx - \sum_{i=1}^{n} t_i Bw_i) + \lambda h = \lambda Lv - Lx ,$$

ce qui implique, en multipliant scalairement par Px et en utilisant (3.9.) et (3.10.),

$$\gamma^{-1}|Bx|^2 - \sum_{i=1}^{n} t_i c(w_i) + (h,x) \leq |Lv|\ |(I-P)x| + (\lambda\theta)^{-1}|Lx|^2 .$$

Par ailleurs, on a, par (3.11.) projetée sur $\text{Im}\, P$ et $\text{Ker}\, P$,

$$|Lx| \leq \lambda(|Bx| + |f|),\quad |(I-P)x| \leq \lambda|K|\ (|Bx| + |f|) ,$$

et dès lors, pour tous les $h \in H$ de norme suffisamment petite,

$$(h,x) \leq (\theta^{-1} - \gamma^{-1})\ |Bx|^2 + c'(h)|Bx| + c''(h).$$

Comme $\theta^{-1} < \gamma^{-1}$, cela entraıne

$$(h,x) \leq c'''(h)$$

et dès lors $|x| \leq R$ pour un certain $R > 0$ par le Théorème de Banach-Steinhaus.

Bibliographie

1 H. Brézis and F.E. Browder, Nonlinear integral equations and systems of Hammerstein type, Advances in Math. 18 (1975) 115-147.

2 H. Brézis and L. Nirenberg, Characterization of the ranges of some

nonlinear operators and applications to boundary value problems, Ann. Sc.Norm.Sup. Pisa (4)5 (1978) 225-326.

3 L. Cesari and R. Kannan, An abstract existence theorem at resonance, Proc. Amer. Math. Soc. 63 (1977) 221-225.

4 D.G. De Figueiredo and C.P. Gupta, Non-linear integral equations of Hammerstein type with indefinite linear Kernel in a Hilbert Space, Indag. Math. 34 (1972) 335-344.

5 J. Mawhin and M. Willem, Compact perurbations of some nonlinear Hammerstein equations, Rivista. Mat. Univ. Parma, (4) 5 (1979) 199-213.

6 J. Mawhin and M. Willem, Range of nonlinear perturbations of linear operators with an infinite dimensional Kernel, Proceed. Conf. Ordin. and Partial Diff. Equ. Dundee, 1978, à paraître.

Jean MAWHIN

Institut Mathématique
Université Catholique de Louvain
2, Chemin du Cyclotron
B-1348 LOUVAIN-LA-NEUVE

BELGIQUE

JEAN MAWHIN

Solutions périodiques d'équations différentielles ordinaires à croissance unilatéralement bornées

1. INTRODUCTION

Ce travail résume l'essentiel des résultats obtenus par l'auteur et W. WALTER [4] et ensuite par J.P. GOSSEZ [2] et relatifs à l'existence de solutions périodiques d'équations différentielles ordinaires dans $\mathbb{R}^n$ à croissance unilatéralement bornée.
Nous résumerons ici la version simplifiée développée dans la monographie [3] à laquelle le lecteur pourra se référer pour de plus amples détails.

Si $I = [0,1]$ et si $f : I \times \mathbb{R}^n \to \mathbb{R}^n$ vérifie les conditions de Carathéodory, on s'intéresse à l'existence des solutions du problème

$$x'(t) = f(t,x(t)), \quad t \in I \tag{1}$$

$$x(0) = x(1) \tag{2}$$

qui seront appelées solutions 1-périodiques de (1).
L'outil utilisé sera une formulation du degré de Leray-Schauder pour certaines perturbations non-linéaires d'opérateurs linéaires de Fredholm, dont nous allons rappeler ici quelques éléments, renvoyant le lecteur aux monographies [1] ou [3] pour plus de détails.

2. ÉLÉMENTS DE THÉORIE DU DEGRÉ DE COINCIDENCE

Soient X et Z des espaces vectoriels normés réels, $\Omega \subset X$ un ouvert borné et $L : \text{dom } L \subset X \to Z$ une application linéaire de Fredholm d'indice zéro, c'est-à-dire telle que $\text{Im } L$ soit fermée et

$$\dim \ker L = \text{codim Im } L < \infty.$$

Soit $\mathscr{C}_L(\Omega)$ la collection des applications

$$F : \text{dom } L \cap \bar{\Omega} \to Z$$

telles que $0 \notin F(\text{dom } L \cap \partial\Omega)$ et qui sont de la forme

$$F = L+G$$

où $G : \bar{\Omega} \to Z$ est L-compacte, c'est-à-dire telle que $\Pi G : \bar{\Omega} \to \text{coker } L$ et $KG : \bar{\Omega} \to X$ soient compactes, avec $\Pi : Z \to \text{coker } L$ la surjection canonique et K un inverse généralisé quelconque de L.

Si $F \in \mathscr{C}_L(\Omega)$, on peut définir, à partir du degré de Leray-Schauder, un entier $D_L(F,\Omega)$, appelé le degré de F relativement à L dans Ω ou encore le degré de coïncidence de L et -G dans Ω, qui conserve les propriétés essentielles du degré de Leray-Schauder (cf. [1] ou [3]).

En particulier, soit $X = C(I,\mathbb{R}^n)$, $Z = L^1(I,\mathbb{R}^n)$, $\text{dom } L = \{x \in X :$ x est absolument continue sur I et $x(0) = x(1)\}$,
$L : x \to x'$, $V \in C^1(\mathbb{R}^n,\mathbb{R})$ telle que $V'(x) \neq 0$ pour tout x tel que $|x| \geq r$, $r > 0$ fixé. Définissons $G : X \to Z$ par

$$G(x)(t) = V'(x(t)), \quad t \in I \tag{3}$$

On a (cf. [4]) la

PROPOSITION 1 : Dans les conditions ci-dessus,

$$F = L+G \in \mathscr{C}_L(B(r)) \text{ et}$$

$$|D_L(F,B(r))| = |d(V',B(r),0| \tag{4}$$

où le second membre de (4) est le degré de Brouwer de V' par rapport à $B(r)$ et 0, $B(r)$ désignant, selon le cas, la boule ouverte de centre 0 et de rayon r dans X ou $\mathbb{R}^n$.

Rappelons enfin un résultat de KRASNOSEL'SKII (cf. [3]) qui assure que si $V'(x) \neq 0$ pour tout $|x| \geq r$ et si

$$V(x) \to +\infty \text{ si } |x| \to \infty ,$$

alors, pour tout $\rho \geq r$,

$$d(V', B(\rho), 0) = +1.$$

3. LE THÉORÈME PRINCIPAL

Le résultat suivant est la formulation donnée en [3] d'un théorème dû à J.P. Gossez [2]. On désigne par $(,)$ le produit scalaire dans $\mathbb{R}^n$.

<u>THEOREME 1</u> : Supposons satisfaites les conditions suivantes :

(i) il existe $V \in C^1(\mathbb{R}^n, \mathbb{R}_+)$ telle que

$$V(x) \to +\infty \text{ si } |x| \to \infty$$

et il existe $a \in L^1(I, \mathbb{R}_+)$ telle que

$$(V'(x), f(t,x)) \leq a(t) \tag{5}$$

pour tout $x \in \mathbb{R}^n$ et presque tout $t \in I$.

(ii) il existe $r > 0$ et $W \in C^1(\mathbb{R}^n \setminus B(r), \mathbb{R})$ tels que, pour $|x| \geq r$

$$(V'(x), W'(x)) > 0 \qquad (6)$$

et tels que

$$\int_I (W'(x(t)), f(t,x(t)))dt \leq 0 \qquad (7)$$

pour tous les $x \in \text{dom } L$ pour lesquels $\min_{t \in I} |x(t)| \geq r$.

Alors le problème (1) - (2) possède au moins une solution.

DEMONSTRATION : Nous en donnerons seulement les grandes lignes. On définit $N : X \to Z$ par

$$(Nx)(t) = f(t,x(t)), \quad t \in I$$

et on considère l'homotopie

$$L - \lambda N + (1-\lambda)G, \qquad \lambda \in [0,1] \qquad (8)$$

où G est défini par (3). Le problème différentiel relatif à (8) est donné par

$$x'(t) = -(1-\lambda)V'(x(t)) + \lambda f(t,x(t)), \quad \lambda \in [0,1]$$
$$x(0) = x(1) \qquad (9)$$

et on va démontrer que les solutions éventuelles de (9) sont bornées a priori indépendamment de $\lambda \in [0,1[$. S'il n'en est pas ainsi, il existe une suite (λ_n) dans $[0,1[$ et une suite (x_n) dans dom L telle que

$$\max_{t \in I} |x_n(t)| \geq n$$

et

$$x_n'(t) = -(1-\lambda_n)V'(x_n(t)) + \lambda_n f(t,x_n(t))$$
$$x_n(0) = x_n(1). \tag{10}$$

Des manipulations sur (10) et la condition (i) impliquent que

$$\frac{d}{dt} V(x_n(t)) \leq a(t), \quad t \in I$$

et dès lors

$$\max_{t\in I} V(x_n(t)) \leq \min_{t\in I} V(x_n(t)) + \int_I |a| .$$

On en déduit aisément que

$$\min_{t\in I} | x_n(t) | \to \infty \text{ si } n \to \infty .$$

D'ailleurs, pour tout $n \in \mathbb{N}^*$ et presque tout $t \in I$,

$$\frac{d}{dt} W(x_n(t)) = -(1-\lambda_n)(V'(x_n(t)),W'(x_n(t)))$$
$$+ \lambda_n(f(t,x_n(t)),W'(x_n(t)))$$

ce qui implique, par (ii), pour n suffisamment grand,

$$0 = - (1-\lambda_n) \int_I (V'(x_n(t)),W'(x_n(t))dt$$
$$+ \lambda_n \int_I (f(t,x_n(t)),W'(x_n(t))dt < 0,$$

une contradiction. Par l'invariance du degré de coïncidence par rapport à l'homotopie ci-dessus et les résultats rappelés à la section 2, on trouve alors, avec $\rho \geq r$ suffisamment grand,

$$|D_L(L-N,B(\rho))| = |D_L(L+G,B(\rho))| = |d(V',B(\rho),0)| = 1,$$

d'où l'existence d'un $x \in \text{dom } L$ tel que $Lx - Nx = 0$, c'est-à-dire d'une solution 1-périodique de (1).

Remarque : Le même résultat reste valable si on remplace $\leq$ dans (5) et (7) par $\geq$. Il suffit de faire la substitution $t \to 1-t$ pour se ramener au Théorème 1.

4. QUELQUES APPLICATIONS

Donnons quelques applications du Théorème 1 obtenues par un choix judicieux de V et W.

COROLLAIRE 1 (GOSSEZ [2]) : Si $n = 1$ et si, pour presque tout $t \in I$, $f(t,x)$ est décroissante en x, alors le problème (1) - (2) a une solution si et seulement s'il existe $y \in L^{\infty}(I,\mathbb{R})$ tel que

$$\int_I f(t,y(t))dt = 0$$

DEMONSTRATION : La condition nécessaire s'obtient en prenant $y = x$. La condition suffisante se déduit du Théorème 1 avec $a(t) = |f(t,0)|$,

$$V(x) = \frac{1}{2} \int_0^{x^2} du/(u^{1/2}+1), \quad r = |y|_{\infty} \quad ,$$

$$W(x) = \frac{1}{2} \int_{r^2}^{x^2} u^{-1/2} du.$$

COROLLAIRE 2 (Mawhin-Walter [4]) : S'il existe $V \in C^1(\mathbb{R}^n, \mathbb{R}_+)$ vérifiant la condition (i) du Théorème 1 et si

$$\int_I \limsup_{|x| \to \infty} (V'(x), f(t,x))dt < 0 , \qquad (11)$$

alors le problème (1) - (2) possède une solution.

DEMONSTRATION : Par le lemme de Fatou et (11), il existe $r > 0$ tel que

$$V'(x) \neq 0$$

pour $|x| \geq r$. Un raisonnement par l'absurde utilisant de nouveau le lemme de Fatou montre alors que $W = V$ satisfait à la condition (ii) du Théorème 1.

Pour de nombreuses autres applications, on pourra consulter [2] , [3] et [4] .

Bibliographie :

1 R.E. Gaines and J. Mawhin, "Coïncidence degree and nonlinear differential equations", Lecture Notes in Math. n°568, Springer, Berlin, 1977.

2 J.P. Gossez, Existence of periodic solutions for some first order ordinary differential equations, Equadiff 78, Firenze, 1978, 361-379.

3 J. Mawhin, "Topological degree methods in nonlinear boundary value problems", CBMS Regional Confer. Series in Math, n°40, Amer. Math. Soc., Providence, 1979.

4 J. Mawhin and W. Walter, Periodic solutions of ordinary differential equations with one-sided growth restrictions, Proc. Royal Soc. Edinburgh 82 A (1978), 95-106.

Jean MAWHIN

Institut Mathématique
Université Catholique de Louvain
2, Chemin du Cyclotron
B-1348 LOUVAIN-LA-NEUVE
BELGIQUE

JEAN MAWHIN

Problèmes résonnants et non résonnants pour une équation des télégraphistes non linéaires

1. INTRODUCTION

Ce travail résume certains résultats obtenus en collaboration avec S. FUČÍK [2] et relatifs à l'existence de solutions périodiques généralisées pour des équations des télégraphistes non-linéaires du type

$$\beta u_t + u_{tt} - u_{xx} - \mu u^+ + \nu u^- + \psi(u) = h(t,x) \qquad (1)$$

où $\beta \neq 0$, μ et ν sont des réels, $\psi : \mathbb{R} \to \mathbb{R}$ est continue et bornée et $h \in L^2(I^2)$ avec $I = [0,2\pi]$. Le travail [2] , qui continue [3] , contient comme ce dernier des résultats relatifs aux cas résonnants et non-résonnants. Nous nous limiterons ici à un seul d'entre eux, renvoyant à [2] et [3] pour les autres, ainsi qu'au travail [1] de Brézis et Nirenberg.

Rappelons qu'une solution périodique généralisée de (1) (en abrégé SPG) est un élément $u \in L^2(I^2) = H$, dont le produit scalaire usuel sera désigné par (,) , tel que, pour toutes les fonctions $v \in C^2(I^2)$ 2π-périodiques en t et x, on ait

$$(u, -\beta v_t + v_{tt} - v_{xx}) = (\mu u^+ - \nu u^- - \psi(u) + h, v)$$

2.

LE CAS OU $\mu = \nu = \lambda$ ET $\psi = 0$. : Par des méthodes classiques utilisant les développements en séries de Fourier, et avec les notations usuelles pour les espaces de Sobolev, on démontre le résultat suivant (voir [2] et [3] pour plus de détails).

PROPOSITION 1. : L'équation

$$\beta u_t + u_{tt} - u_{xx} - \lambda u = h(t,x) \qquad (2)$$

où $\beta \neq 0$ et λ sont des réels, possède une SPG pour tout $h \in H$ si et seulement si

$$\lambda \neq m^2 , \qquad m \in \mathbb{N} ,$$

auquel cas la solution est unique et définit un opérateur linéaire compact

$$T_\lambda : H \to H , \qquad h \mapsto u ,$$

tel que

$$T_\lambda(H) \subset W^{1,2}(I^2) \cap C(I^2)$$

$$T_\lambda(W^{k,2}(I^2)) \subset W^{k+1,2}(I^2) , \quad k \geq 1.$$

Si $\lambda = q^2$ pour un certain entier $q \geq 0$, alors (2) possède une solution si et seulement si

$$h \in H_q = \{h \in H : \int_0^{2\pi} \int_0^{2\pi} h(t,x)e^{iqx} \, dx \, dt = 0\} ,$$

auquel cas il existe une SPG unique $u \in H_q$ définissant un opérateur linéaire compact

$$\tilde{T}_q : H_q \to H_q , \qquad h \mapsto u ,$$

tel que

$$\tilde{T}_q(H_q) \subset W^{1,2}(I^2) \cap H_q \cap C(I^2)$$

$$\tilde{T}_q(H_q \cap W^{k,2}(I^2)) \subset H_q \cap W^{k+1,2}(I^2), \quad k \geq 1.$$

LE CAS OU $\Psi = 0$. : Considérons tout d'abord l'équation (1) où $\Psi = 0$ et $h = 0$, c'est-à-dire

$$\beta u_t + u_{tt} - u_{xx} - \mu u^+ + \nu u^- = 0. \tag{3}$$

ou encore

$$\beta u_t + u_{tt} - u_{xx} = \mu u^+ - \nu u^-.$$

Par la Proposition 1, et puisque u^+ et u^- sont dans H si $u \in H$, toute SPG de (3) appartient à $W^{1,2}(I^2)$, ce qui implique que $\mu u^+ - \nu u^- \in W^{1,2}(I^2)$. Appliquant une nouvelle fois cette proposition, on trouve que $u \in W^{2,2}(I^2)$ et dès lors, par la formule de Green, u vérifie (3) presque partout sur I^2. On déduit alors de (3) que

$$\beta \int_{I^2} u_t^2 = 0$$

et dès lors u est indépendante de t et vérifie, au sens de Caratheodory, le problème (avec $u'' = d^2u/dx^2$)

$$\begin{aligned} &- u'' = \mu u^+ - \nu u^- \\ &u(0) - u(2\pi) = u'(0) - u'(2\pi) = 0. \end{aligned} \tag{4}$$

Mais toute solution, au sens de Caratheodory, de (4) est une solution classique et une analyse élémentaire de (4) montre que (4) possède une solution non triviale si et seulement si l'une des conditions suivantes est satisfaite :

$$\mu = 0 , \quad \nu \in \mathbb{R} \tag{5}$$

$$\nu = 0 , \quad \mu \in \mathbb{R} \tag{6}$$

$$\mu > 0 , \quad \nu > 0 \text{ et } k(\mu^{-1/2}+\nu^{-1/2}) = 2 , \quad k \in \mathbb{N}\setminus\{0\}. \tag{7}$$

Nous avons donc la

PROPOSITION 2. : *L'équation* (3) *possède une* SPG *non triviale si et seulement si l'une des conditions* (5), (6) *ou* (7) *est vérifiée.*

Passons maintenant au cas où $\Psi = 0$ et notons tout d'abord que si $\mu > 0$ et $\nu < 0$ (resp. $\mu < 0$ et $\nu > 0$), il existe $h \in L^2(I^2)$ tel que l'équation correspondante

$$\beta u_t + u_{tt} - u_{xx} - \mu u^+ + \nu u^- = h(t,x) \tag{8}$$

n'a pas de SPG. En effet si, pour fixer les idées, $\mu > 0$ et $\nu < 0$, alors, en prenant $h(t,x) = 1$, on montre comme ci-dessus que toute SPG de (8) avec $h(t,x) = 1$ ne dépend pas de t et vérifie

$$\begin{aligned} & u'' + \mu u^+ - \nu u^- + 1 = 0 \\ & u(0) - u(2\pi) = u'(0) - u'(2\pi) = 0 \end{aligned} \tag{9}$$

et donc

$$u'' + 1 \leq 0$$

ce qui est impossible pour une solution 2π-périodique. On ignore actuellement la caractérisation de l'ensemble des (μ,ν) pour lesquels (8) possède une SPG quel que soit $h \in L^2(I^2)$; la remarque ci-dessus montre que cet ensemble est contenu dans

$$M = \{(\mu,\nu) \in \mathbb{R}^2 : \mu\nu \geq 0\}$$

et on va montrer au paragraphe suivant que cet ensemble contient l'ensemble

$$M_o = M \setminus \{(\mu,\nu) \in \mathbb{R}^2 : (5) \text{ ou } (6) \text{ ou } (7) \text{ est satisfaite}\}$$

4.

LE CAS GENERAL : On vérifie sans peine que M_o est une partie ouverte de $\mathbb{R}^2$. Soit $(\mu,\nu) \in M_o$; il existe donc $\varepsilon \in]0,1[$ tel que $(\mu-\varepsilon,\nu-\varepsilon)$ appartienne à la même composante connexe de M_o que (μ,ν). On écrit l'équation(1) sous la forme équivalente.

$$\beta u_t + u_{tt} - u_{xx} - \varepsilon u = (\mu-\varepsilon)u^+ - (\nu-\varepsilon)u^- + \psi(u) + h(t,x) \qquad (10)$$

et dès lors, par les résultats de la Proposition 1, les SPG de (10), et donc de (1) sont les solutions dans H de

$$u = T_\varepsilon [(\mu-\varepsilon)u^+ - (\nu-\varepsilon)u^- + \psi(u) + h] \qquad (11)$$

Introduisons la famille auxiliaire

$$u - R_\varepsilon u = S(u,\lambda), \qquad \lambda \in [0,1]$$

où

$$R_\varepsilon u = T_\varepsilon [(\mu-\varepsilon)u^+ - (\nu-\varepsilon)u^-]$$

$$S(u,\lambda) = \lambda T_\varepsilon (h-\psi(u)).$$

On vérifie sans peine que R_ε et S sont compactes sur tout borné et dès lors, comme

$$u - R_\varepsilon u = 0 \iff u = 0$$

un argument traditionnel de compacité implique l'existence de $c > 0$ tel que, pour tout $u \in H$,

$$|u - R_\varepsilon u| \geq c|u| .$$

D'ailleurs, il existe $C > 0$ tel que, pour tout $u \in H$, et tout $\lambda \in [0,1]$,

$$|S(u,\lambda)| \leq C$$

et dès lors, si $R > C/c$, on aura, pour $|u| = R$ et $\lambda \in [0,1]$,

$$|u - R_\varepsilon u - S(u,\lambda)| \geq cR - C > 0 ,$$

ce qui implique l'existence du degré de Leray-Schauder

$$d[I - R_\varepsilon - S(.,\lambda), B(R), 0]$$

et son indépendance par rapport à $\lambda \in [0,1]$. Pour calculer sa valeur en $\lambda = 0$, c'est-à-dire

$$d[I-R_\varepsilon,B(R),0] ,$$

on note que la composante connexe de M_o contenant $(\mu-\varepsilon,\nu-\varepsilon)$ et (μ,ν) contient aussi un élément de la forme (λ,λ), avec $\lambda \neq m^2$ $(m \in \mathbb{N})$ et dès lors, en vertu de l'invariance du degré de Leray-Schauder par homotopie,

$$d[I-R_\varepsilon,B(R),0] = d[I-(\lambda-\varepsilon)T_\varepsilon,B(R),0] \tag{12}$$

Comme $u = (\lambda-\varepsilon)T_\varepsilon u$ si et seulement si u est SPG de (2) avec $h = 0$ et que $\lambda \neq m^2$ $(m \in \mathbb{N})$, $\ker(I-(\lambda-\varepsilon)T_\varepsilon) = \{0\}$ et le second membre de (12) est égal à ± 1. Donc,

$$d[I-R_\varepsilon-S(.,1),B(R),0] = \pm 1$$

et, par le théorème d'existence de Leray-Schauder, nous avons démontré le

THEOREME 1. : Si $\beta \neq 0$, $(\mu,\nu) \in M_o$ et $\psi : \mathbb{R} \to \mathbb{R}$ est continue et bornée, alors, pour tout $h \in L^2(I^2)$, l'équation (1) possède au moins une SPG.

Bibliographie

1 H. Brézis et L. Nirenberg, Characterizations of the ranges of some nonlinear operators and applications to boundary value problems, Ann. Sc. Norm. Sup. Pisa (4) 5, (1978), 225-326.

2 S. Fučik and J. Mawhin, Generalized periodic solutions of nonlinear telegraph equations, Nonlinear Analysis 2 (1978), 609-617.

3 J. Mawhin, Periodic solutions of nonlinear telegraph equations, Dynamical Systems, Academic Press Press, New-York, (1977), 193-210.

Jean MAWHIN

Institut Mathématique
Université Catholique de Louvain
2, Chemin du Cyclotron
B-1348 LOUVAIN-LA-NEUVE

BELGIQUE

L A PELETIER

Large time behaviour of solutions of the porous media equation

1. INTRODUCTION

Let Ω be a bounded domain in $\mathbb{R}^n$ with smooth compact boundary $\partial\Omega$, and suppose $u = u(x,t)$ is a solution of the initial-boundary value problem

$$(I)\begin{cases} u_t = \Delta(u^m) \text{ on } Q = \Omega\times(0,\infty) & (1)\\ u = 0 \quad \text{on } S = \partial\Omega\times(0,\infty) & (2)\\ u = u_o \quad \text{on } \bar{\Omega}\times\{0\}, \end{cases}$$

in which $m \geq 1$ and u_o is a given non-negative function.

If m=1, equation (1) is the heat equation and it is well known that under appropriate conditions on u_o, $u(.,t) \to 0$ as $t \to \infty$. More precisely, let $0 < \lambda_o < \lambda_1 \leq \ldots$ be the eigenvalues of $-\Delta$ on Ω with Dirichlet boundary conditions, and let ϕ_o be the eigenfunction corresponding to λ_o, then

$$\|u(.,t)\| \leq K_1 e^{-\lambda_o t} \qquad t \geq 0 \qquad (A_1)$$

$$\|e^{\lambda_o t} u(.,t) - (\phi_o,u_o)\phi_o\| \leq K_2 e^{-(\lambda_1-\lambda_o)t} \qquad t \geq 0. \qquad (B_1)$$

Here $\| \;\|$ and $(\; , \;)$ denote, respectively, the norm and the inner product on $L^2(\Omega)$ and K_1 and K_2 are constants which only depend on u_o.

The object of this talk is to obtain estimates like (A_1) and (B_1) for

---

* The work reported in this lecture has been done jointly with D.G. Aronson, University of Minnesota [1].

solutions of Problem I if $m > 1$. Before tackling this problem in its full generality, we consider a particular class of solutions of equation (1), obtained by separating the variables. This yields solutions of the form

$$\bar{u}(x,t) = (t+\tau)^{-\gamma} f(x),$$

where $\tau > 0$ is an arbitrary constant, $\gamma = 1/(m-1)$ and f a solution of the problem

$$\begin{cases} \Delta(f^m) + \gamma f = 0 & \text{in } \Omega \\ \qquad\qquad f = 0 & \text{on } \partial\Omega. \end{cases} \tag{II}$$

It can be shown that this problem has a unique positive solution for any $\gamma \in (0,\infty)$.

The two main estimates can now be formulated in terms of γ and the function f :

$$\|u(.,t)\|_\infty \leq K_1(1+t)^{-\gamma} \tag{A_m}$$

$$\|(1+t)^\gamma u(.,t) - f\|_\infty \leq K_2(1+t)^{-1}. \tag{B_m}$$

Here $\| \ \|_\infty$ denotes the norm on $L^\infty(\Omega)$ and K_1, K_2 are constants which only depend on u_o, m and Ω. The exponents of $(1+t)$ are both optimal.

2. PRELIMINARIES

It is well known that if $m > 1$, Problem I need not possess a classical solution. Thus we need a weaker notion of solution.

Let Q_T denote the cylinder $\Omega \times [0,T]$ and S_T its lateral boundary $\partial\Omega \times [0,T]$.

DEFINITION : A function $u : \bar{Q}_T \to [0,\infty)$ is called a *weak solution* of Problem I if

(i) for each $(y,t) \in S_T$, $\lim_{(x,t)\to(y,t)} u(x,t) = 0$ where $x \in \Omega$;

(ii) $(u^m)_x$ exists in a distributional sense in Q_T and

$$\iint_{Q_T} \{u^2 + |(u^m)_x|^2\} dxdt < \infty ;$$

(iii) $$\iint_{Q_T} \{\phi_x (u^m)_x - \phi_t u\} dxdt = \int_\Omega \phi(x,0) u_o(x) dx$$

for any $\phi \in C^1(\bar{Q}_T)$ which vanishes on S_T and $\Omega \times \{T\}$.

This definition follows the one given by Oleinik, Kalashnikov and Yui-Lin [2] who considered Probem I in one dimension, and Sabinina [3], who considered the Cauchy Problem in n dimensions.

About the function u_o we make the following hypotheses :

$u_o^m \in C^1(\bar{\Omega})$, $u_o = 0$ on $\partial\Omega$; H1.

$u_o \geq 0$ $(\not\equiv 0)$ in $\bar{\Omega}$. H2.

THEOREM 1 : Suppose u_o satisfies hypotheses H1 and H2. Then Problem I has one and only one weak solution.

THEOREM 2 : Suppose u_o and v_o satisfy hypotheses H1 and H2, and u and v are the weak solutions corresponding to u_o and v_o. Then, $u_o \geq v_o$ in $\bar{\Omega}$ implies $u \geq v$ in $\bar{Q}_T$.

Let $e : \bar{\Omega} \to \mathbb{R}$ be the solution of the problem

$$\begin{cases} -\Delta u = 1 & \text{in } \bar{\Omega} \\ u = 0 & \text{on } \partial\Omega. \end{cases}$$

Then, by the maximum principle $e(x) > 0$ in Ω and $\partial e/\partial n < 0$ on $\partial\Omega$. Define the set

$$K = \{\phi \in C^1(\bar{\Omega}) : \phi \geq ke \text{ in } \bar{\Omega} \text{ for some } k > 0\}.$$

THEOREM 3 : Let f be the positive solution of Problem I. Then $f^m \in K$.

3. THE ESTIMATES

THEOREM 4 : Suppose u_o satisfies hypotheses H1 and H2, and let u be the weak solution of Problem I which corresponds to u_o. Then there exists a constant $\tau_1 > 0$, which only depends on u_o and m, such that

$$u(x,t) \leq (\tau_1+t)^{-\gamma} f(x) \text{ in } \bar{Q}. \tag{3}$$

PROOF : The function

$$v(x,t) = (\tau+t)^{-\gamma} f(x)$$

is a solution of Problem I. By Theorem 3, there exists a $\tau_1 > 0$ such that

$$f(x) \geq \tau_1^{\gamma} u_o(x) \quad \text{in } \bar{\Omega},$$

and hence the result follows from Theorem 2.

Remark : Suppose that u_o satisfies hypothesis H1 but, instead of H2, it satisfies the stronger hypothesis

$$u_o^m \in K. \qquad\qquad H2^*.$$

Then, there exists a $\delta > 0$ such that

$$u_o(x) \geq \delta\, f(x) \text{ in } \bar{\Omega}$$

and hence, by Theorem 2,

$$u(x,t) \geq (\tau_2+t)^{-\gamma}\, f(x) \quad \text{in } \bar{Q}, \tag{4}$$

where $\tau_2 = \delta^{-1/\gamma}$.

It follows from (3) that

$$(1+t)^{\gamma}\, u(x,t) - f(x) \leq \{ (\frac{1+t}{\tau_1+t})^{\gamma} - 1 \} f(x) \quad \text{on } \bar{Q}$$

and from (4) that

$$(1+t)^{\gamma}\, u(x,t) - f(x) \geq \{ (\frac{1+t}{\tau_2+t})^{\gamma} - 1 \} f(x) \quad \text{on } \bar{Q}.$$

Hence

$$|(1+t)^{\gamma}\, u(x,t) - f(x)| \leq M(1+t)^{-1} f(x) \qquad \text{on } \bar{Q},$$

where M depends on τ_1, τ_2, and hence on u_o and m.

In the remainder of this talk we shall show that the positivity hypothesis $H2^*$ may be replaced by H2.

4. POSITIVITY

LEMMA 1 : Let u be the weak solution of Problem I in which u_o satisfies H1 and H2. Then there exists a constant $T > 0$ such that

$$u(x,t) > 0 \text{ for } (x,t) \in \Omega \times [T,\infty).$$

PROOF : Consider the function

$$p(x-\xi,t;a,\tau) = [c(t+\tau)^{-\alpha}\{a^2 - |x-\xi|^2(t+\tau)^{-2\beta}\}]_+,$$

where $[f]_+ = \max\{0,f\}$, $\alpha = n\beta = n\{(m-1)n+2\}^{-1}$, $c = 2m(n+2\gamma)^{-\gamma}$, $\xi \in \mathbb{R}^n$ and $a,\tau \in (0,\infty)$. It can be verified that p is a weak solution of the Cauchy Problem for equation (1).

In view of H2, there exist $\xi_1 \in \mathbb{R}^n$, $a_1,\tau_1 \in (0,\infty)$ such that

$$p_1(x,0) \equiv p(x-\xi_1,0;a_1,\tau_1) \leq u_0(x) \text{ on } \bar{\Omega},$$

and hence, by Theorem 2,

$$p_1(x,t) \leq u(x,t)$$

as long as $\operatorname{supp}(p_1) \subset \bar{\Omega}$. Utilizing the fact that the support of p increases with time, and choosing an appropriate sequence of functions p_i, $i=1,2,\ldots,N$, it is possible to prove the Lemma.

Lemma 1 does not yet imply that $u^m(.,T) \in K$. However, given that $u(x,T) > 0$ in Ω, it is possible to prove in a similar manner, using a different comparison function, that there exists a constant $T' > 0$ such that $u^m(.,T+T') \in K$.

Bibliography

1 D.G. Aronson and L.A. Peletier, Large time behaviour of solutions of the porous media equation. To appear in J. Diff. Equations.

2 O.A. Oleinik, A.S. Kalashnikov and Chzhou Yui-Lin, The Cauchy problem and boundary problems for equations of the type of

unsteady filtration Izv. Akad. Nauk. SSSR Ser. Mat. 22, 667-704 (1958).

E.S. Sabinina, On the Cauchy problem for the equation of nonstationary gas filtration in several space variables. Doklady Akad. Nauk 136, 1034-1037 (1961), (Sov. Mat. 1, 166-169) (1961).

L.A. PELETIER

Mathematisch Instituut
Der Rijksuniversiteit Te Leiden
2300-RA LEIDEN
The NETHERLANDS

L A PELETIER

On an equation arising in the theory of turbulence*

1. INTRODUCTION

Recently, Brauner, Penel and Temam [2] gave a study of the Cauchy problem

$$P(\nu)\begin{cases} u_t + ([u(0,t) - u(x,t)]^2)_{xx} = \nu u_{xx} & \text{in } S_T = \mathbb{R}\times(0,T] \quad (1)\\ u(x,0) = u_o(x) & \text{on } \mathbb{R} \end{cases}$$

in which $T > 0$ and $\nu \geq 0$. This problem arises in a stochastic model for turbulence, based on the Burgers equation

$$u_t + uu_x = \nu u_{xx} ,$$

which was introduced by Burgers as a simplified version of the Navier-Stokes equations [4] . In the context of this model, the variable u in problem $P(\nu)$ represents a covariance, so it is natural to assume that u is positive definite ($u >> 0$).

For future reference we quote a few results of [2] :

(a) If $\nu > 0$, $u_o >> 0$ and $u_o \in H^1(\mathbb{R})$, there exists a unique solution u_ν of Problem $P(\nu)$ in some weak sense. Moreover $u_\nu(\cdot,t) >> 0$ for each $t \in [0,T]$.

(*)The work reported in this lecture is mainly due to C.J. van Duyn, (cf. [3]) University of Leiden, Netherlands.

(b) Let $\{\nu_k\}$ ba a sequence such that $\nu_k \to 0$ as $k \to \infty$. Then $u_{\nu_k} \to u$ as $k \to \infty$ in $L^\infty([0,T]; H^1(\mathbb{R}))$ weak star.

(c) If $\nu \geq 0$, then

$$\| u_\nu \|_{L^2(S_T)} \leq \| u_o \|_{L^2(\mathbb{R})} .$$

Finally, Penel [8] showed :

(d) If $\nu > 0$, $u_o >> 0$ and $u_o \in C^\infty(\mathbb{R})$, then $u_\nu \in C^\infty(\bar{S}_T)$.

In this talk we shall be particularly interested in Problem P(0). Like the porous media equation

$$u_t = (u^m)_{xx} \qquad m > 1 ,$$

equation (1) is degenerate parabolic, and need not possess a classical solution, however smooth we choose u_o. We shall introduce a class of generalized solutions, similar to the one defined by Oleinik, Kalashnikov and Yui-Lin for the porous media equation [7]. For this class of solutions we shall obtain the following results :

1. Existence of a generalized solution;
2. Optimal regularity of generalized solutions;
3. If $1 - Ax^2 \leq u_o(x) \leq 1$ for $|x| \leq \ell$, $(A, \ell > 0)$, then there exists a constant $T_o > 0$ such that $u(0,t) = 1$ for $0 \leq t \leq T_o$.

2. EXISTENCE

<u>DEFINITION</u> : A function $u : \bar{S}_T \to \mathbb{R}$ is called a generalized solution of Problem P(0) if

(i) u is continuous and bounded on $\bar{S}_T$;

(ii) $u(.,t) >> 0$ for all $t \in [0,T]$;

(iii) $[u(0,t) - u(x,t)]^2$ has a bounded generalized derivative with respect to x in S_T ;

(iv) u satisfies the identity

$$\iint_{S_T} \{\phi_x (v^2)_x + \phi_t u\} dx\, dt = - \int_{-\infty}^{\infty} \phi(x,0) u_o(x) dx$$

in which $v(x,t) = u(0,t) - u(x,t)$; for all $\phi \in C^1(\overline{S}_T)$ such that $\phi = 0$ for large $|x|$ and for $t = T$.

<u>THEOREM 1.</u> : Let $u_o \in L^2(\mathbb{R})$ be uniformly Lipschitz continuous on $\mathbb{R}$ and $u_o >> 0$. Then there exists a generalized solution of Problem P(0).

<u>Proof</u> : We shall seek the solution u of Problem P(0) as the limit of a sequence of smooth solutions of the regularized problem $P(\nu)$. Let u_n be the solution of Problem P(1/n), with initial value

$$u_{o,n}(x) = \int_{-\infty}^{\infty} \rho_{1/n}(x-y)\ u_o(y)\, dy$$

in which $\rho_{1/n}(x) = \pi^{-1/2} n\, \rho(nx)$ and $\rho(x) = e^{-x^2}$. It can be shown by means of a maximum principle argument, originally due to Bernstein [1] that

$$|u_n(x',t) - u_n(x'',t)| \le K_1 |x'-x''| \qquad (2)$$

for all $x', x'' \in \mathbb{R}$ and $t \in [0,T]$ and hence, by a result due to Gilding [5] that

$$|u_n(x,t') - u_n(x,t'')| \le K_2 |t'-t''|^{1/2} \qquad (3)$$

for all $x \in \mathbb{R}$ and $t',t'' \in [0,T]$. Thus the sequence $\{u_n\}$ is equicontinuous in $\overline{S}_T$ whence there exists a subsequence $\{u_{n_k}\}$ which converges to a function

$u \in C(\bar{S}_T)$ uniformly on compact subsets of $\bar{S}_T$.

It is readily shown that u has the properties (i), (iii) and (iv) of a generalized solution. To prove (ii) we observe that since $u_{n_k}(.,t) >> 0$

$$\sum_{i,j=1}^{N} u_{n_k}(x_i - x_j, t) \ \alpha_i \bar{\alpha}_j \ \geq 0 \tag{4}$$

for any finite set of points $x_1,\ldots,x_N$ and complex numbers $\alpha_1,\ldots,\alpha_N$. Since the points $x_1,\ldots,x_N$ are contained in a compact set of $\mathbb{R}$, it follows that (4) continues to hold for u. Thus $u(.,t) >> 0$ for any $t \in [0,T]$. This completes the proof.

3. REGULARITY

<u>THEOREM 2</u> : The generalized solution u of Problem P(0), constructed in Theorem 1 has the following properties :

(i) For any $t \in [0,T]$,

$$u(.,t) \in L^2(\mathbb{R}) \text{ and } \|u(.,t)\|_{L^2(\mathbb{R})} \leq \|u_o\|_{L^2(\mathbb{R})} .$$

(ii) $u(x,t)$ is uniformly Lipschitz continuous with respect to x and uniformly Hölder continuous (exponent 1/2) with respect to t on $\bar{S}_T$.

(iii) u is a classical solution of equation (1) in a neighbourhood of any point $(x_o,t_o) \in S_T$ with $x_o \neq 0$.

(iv) Given any $\lambda > 0$, $\partial v^{1+\lambda}/\partial x$ exists and is continuous in S_T . Moreover, $\partial v^{1+\lambda}(0,t)/\partial x = 0$ for all $t \in (0,T]$.

Proof : (i) It follows from [2c] that for each $t \in [0,T]$ and all $n \geq 1$

$$\|u_n(\cdot,t)\|_{L^2(\mathbb{R})} \leq \|u_{o,n}\|_{L^2(\mathbb{R})} .$$

Moreover since

$$\|u_{o,n}\|_{L^2(\mathbb{R})} \leq \|u_o\|_{L^2(\mathbb{R})} \qquad n \geq 1$$

the result follows at once.

(ii) This property follows directly from (2) and (3).

(iii) Let $(x_o,t_o) \in S_T$. Then, because $u(\cdot,t_o) >> 0$,

$$u(x_o,t_o) < u(0,t_o) \text{ if } x_o \neq 0 .$$

Thus, there exists an $\varepsilon > 0$ and a neighbourhood N of (x_o,t_o) such that

$$u_n(0,t) - u_n(x,t) > \varepsilon \quad \text{in } N$$

for n large enough. This means that equation (1) is uniformly parabolic in N for $\nu \geq 0$. The result now follows from standard theory.

(iv) Let $Q_\delta = \{(x,t) : |x| < \delta , 0 < t \leq T\}$. Then it can be shown that for any $\lambda > 0$:

$$0 \leq v^{1+\lambda}(x,t) \leq (1+\lambda)C\delta^\lambda \; |x|$$

where C is a positive constant. It follows that $\partial v^{1+\lambda}(0,t)/\partial x = 0$ for all $t \in (0,T]$. The continuity of $\partial v^{1+\lambda}/\partial x$ follows similary.

To show that the regularity result obtained in Theorem 2 is optimal, we consider a particular solution. To begin with we consider the half-space problem :

$$(I^+)\begin{cases} u_t + (v^2)_{xx} = 0 & x > 0,\ t > 0 \\ u(0,t) = U(t+1)^{\gamma} & t \geq 0 \\ u(x,0) = u_o(x) & x \geq 0\ , \end{cases}$$

where U is an arbitrary positive number, and γ and u_o are quantities which will be determined later. We look for a solution of the form

$$u(x,t) = U(t+1)^{\gamma}\ f(\eta) \qquad \eta = x(t+1)^{-\beta}\ ;$$

which reduces Problem (I^+) to :

$$\begin{cases} ([f(0) - f(\eta)]^2)'' - r\,\eta\, f' + s\, f = 0,\ \eta > 0 & (5) \\ \\ f(0) = 1,\quad f(\infty) = 0\ , \end{cases}$$

where $r = \beta/U$ and $s = (2\beta-1)/U$. This problem has a solution f^+ if we choose $\beta = \gamma = 1/3$. Moreover $(f^+)'(0) = -1/\sqrt{6}$. By symmetry we can define a function f^- on $(-\infty,0)$ such that f^- satisfies (5) for $\eta < 0$ and $f^-(0) = 1$, $f^-(-\infty) = 0$. Finally we define

$$f(\eta) = \begin{cases} f^+(\eta) & \text{when} \quad \eta \geq 0 \\ \\ f^-(\eta) & \text{when} \quad \eta \leq 0\ . \end{cases}$$

Then it can be shown that the function

$$\overline{u}(x,t) = U(t+1)^{-1/3} f(x(t+1)^{-1/3})$$

is a generalized solution of Problem P(0) for an appropriate choice of u_o . Clearly $\overline{u}(.,t) \in \mathrm{Lip}$ for any $t \in [0,T]$.

4. A GEOMETRICAL PROPERTY

Consider the problem

$$\begin{cases} u_t = (u^m)_{xx} & \text{in } S_T \\ u(x,0) = u_o(x) & \text{on } \mathbb{R} \end{cases}$$

and suppose that

$$0 \le u_o(x) \le Ax^2 \ ,$$

then it was shown by Kalashnikov [6] that there exists a $T_o > 0$ such that

$$u(0,t) = 0 \quad \text{for } t \in [0,T_o] \ .$$

We prove an analogous result for Problem P(0).

THEOREM 3. : Let u be the generalized solution of Problem P(0) constructed in Theorem 1, and let u_o satisfy

$$1 - Ax^2 \le u_o(x) \le 1 \qquad \text{for } |x| \le \ell \ ,$$

where A and ℓ are positive constants. Then

$$u(0,t) = 1 \qquad \text{for } t \in [0,T_o] \ ,$$

where $T_o = \frac{1}{12} \min \{A^{-1}, 1/2\ \ell^2\}$.

The proof relies on the construction of a rather involved comparison function. For the details we refer to [3].

Bibliography :

1 D.G. Aronson, Regularity properties of flows through porous media, Siam J. Appl. Math. 17, 461-467 (1969).

2 C.M. Brauner, P. Penel & R. Temam, Sur une équation d'évolution non linéaire liée à la théorie de la turbulence, Annali della Scuola Norm. Sup. di Pisa IV, 101-128 (1977).

3 Duyn, C.J. van, Regularity properties of solutions of an equation arising in the theory of turbulence, J. Differential Equations, 33 (1979) 226-238.

4 U. Frisch, M. Lesieur & A. Brissaud, A Markovian random coupling model for turbulence, J. Fluid Mech. 65,145-152 (1974).

5 B.H. Gilding, Hölder continuity of solutions of parabolic equations, J. London Math. Soc. (2), 13, 103-106 (1976).

6 A.S. Kalashnikov, The occurrence of singularities in solutions of the non-steady seepage equation, USSR Computational Math. and Math. Phys. 7, 269-275 (1967).

7 O.A. Oleinik, A.S. Kalashnikov & Chzhou Yui-Lin, The Cauchy problem and boundary problems for equations of the type of unsteady filtration, Izv. Akad. Nauk. SSSR Ser. Mat. 22, 667-704 (1958).

8 P. Penel, Sur une équation d'évolution non lineaire liée à la théorie de la turbulence : II, Existence et régularité de solutions "fortes", to appear.

L.A. PELETIER

Mathematisch Instituut
Der Rijksuniversiteit Te Leiden
2300-RA LEIDEN
The NETHERLANDS

MICHAEL RENARDY

Bifurcation of singular and transient solutions. Spatially nonperiodic patterns for chemical reaction models in infinitely extended domains

ABSTRACT : We discuss bifurcation of trajectories connecting saddle points from stationary solutions. The results are applied to models for chemical reactions.

0. INTRODUCTION

We shall consider a differential equation

$$\frac{du}{dx} = \dot{u} = A(\mu)u + B(\mu,u) \qquad (0.1.)$$

where μ is a real parameter and u is in a Banach space Y. $A(\mu)$ is an (in general unbounded) linear operator in Y, which satisfies certain semi-group conditions specified later, and B is a smooth nonlinear term satisfying $\|B(\mu,u)\| = O(\|u\|^2)$. We assume (0.1.) is reversible in the sense of Moser [20], i.e. there exists a linear mapping $R \in \mathcal{L}(Y)$ such that $R^2 = id$, $RA(\mu) = -A(\mu)R$ and $B(\mu,Ru) = -RB(\mu,u)$. Furthermore, we shall assume that for $\mu < 0$ the spectrum of $A(\mu)$ is bounded away from the imaginary axis, whereas at $\mu = 0$ a pair of eigenvalues passes through zero and becomes imaginary for $\mu > 0$. It can then be proved [15], [16] that for each $\mu \gtrsim 0$ there exists a one parameter family of periodic orbits centered at the origin. It was further shown by Kirchgässner and Scheurle in [15], [16] that under certain assumptions bounded nonperiodic solutions can be constructed as a limit of these periodic solutions, the convergence being uniform in bounded intervals. Kirchgässner and Scheurle call these solutions "singular".
In this paper a direct approach for the construction of bifurcating singular solutions is given, which does not use approximation by periodic orbits.

On the contrary we make explicit use of the fact that singular solutions are isolated in a suitable space of functions. We introduce a bifurcation parameter ε and investigate (0.1.) in lowest order in ε, where singular solutions can be found explicitly. At this point it is essential that besides a scaling of u and μ by appropriate powers of ε we must also scale the independent variable x. This procedure has already been used in earlier work by other authors, e.g. in [9], [28]. We then discuss the kernel and codimension of the linearization at these singular solutions. Using methods similar to those employed in [26], this can be done without solving the linearized equations explicity. We must only consider the asymptotic behaviour of solutions for $x \to \pm\infty$, where the singular solutions approach saddle points. Finally a generalized form of the implicit function theorem, which has been developed in [25] for a different purpose, is used to prove the existence of singular solutions for $\varepsilon \neq 0$.

In order not to complicate our arguments by too many purely technical complexities, we shall confine ourselves in this "general" part to "generic" cases. As we shall see later in the examples, however, our methods can in many cases still be applied also under nongeneric conditions, but modifications, e.g. in the choice of scaling factors, are necessary. Under generic hypotheses two cases must be distinguished, which correspond to the two kinds of reversibility considered in [15]. In the first case a twosided branch of stationary solutions emanates from $u = 0$. The linearized spectrum at this bifurcating fixed point contains no imaginary eigenvalues for $\mu > 0$, whereas two imaginary eigenvalues occur for $\mu < 0$. We find that for $\mu < 0$ the fixed point 0 and for $\mu > 0$ the new fixed point are connected to themselves by a biasymptotic trajectory. In the second case we have a onesided bifurcation of two fixed points. These are connected to each other by two trajectories, if the bifurcation goes to positive values of μ .

In section II we shall discuss a different problem, which can be treated using the same methods. We shall consider a non-reversible equation of the form (0.1.) and assume that one simple eigenvalue of $A(\mu)$ passes through 0 at $\mu = 0$. It is well known that under generic hypotheses a new branch of stationary solutions bifurcates [5] . We prove the existence of solutions approaching one fixed point as $x \to \infty$ and the other one as $x \to -\infty$. We call these solutions "transient".

Our results are related to the work of Kopell and Howard [18] . They discuss bifurcation of trajectories connecting critical points using the center manifold theorem and transversality arguments. Our results in §6 can also be obtained by their method, but our proof is different. A proof for the existence of transient solutions which is more similar to ours is given by Nicolaenko in [30] .

A physical example for the occurrence of singular solutions, which originally motivated the work of Kirchgässner and Scheurle, is provided by stationary solutions of the Bénard problem. Moreover, some equations are known for which singular solutions (called "solitons") have been found explicitly. These equations are relevant e.g. in the theory of Josephson junctions and for the phenomenon of self-induced transparency in nonlinear optics [3] . We shall in section III apply our results to models for chemical reactions. These are in general equations of the form

$$\frac{\partial^2 u_i}{\partial x^2} = f_i(u_1,\ldots,u_n) + \alpha_i \frac{\partial u_i}{\partial t} \qquad (0.2.)$$

where the f_i are functions of polynomial type. Clearly (0.2.) is reversible with respect to the variable x. We shall discuss three different types of spatially nonperiodic solutions. In §7 we confine ourselves to time (i.e. t) independent bifurcating singular solutions. Since these bifurcations are structurally unstable, we have to consider imperfect bifurcations as well. We find time-independent solutions approaching the same constant values as $x \to \pm\infty$. For a special example, where all stationary solutions could be given exactly, solutions of the type considered here were found in [11]. In §8 we shall assume that in the x-independent equation (0.2.) there occurs a Hopf bifurcation. This is known to be the case in some often studied models as e.g. the Brusselator [1], [9] and the Field-Noyes model of the Belousov-Zhabotinskii reaction [6], [10] [12], [13], [21], [23]. We find the existence of time-periodic solutions approaching the original stationary state as $x \to \pm\infty$. In §9 we are finally concerned with "transient" solutions of (0.2.), which depend only on the one independent variable $x' = x-\gamma t$, i.e. with travelling waves. In special cases, e.g. if only one species of chemicals is involved, such solutions have already been discussed previously [2], [7], [8], [27].

It is essential in our analysis that in all the cases discussed the asymptotic behaviour as $x \to \pm\infty$ is constant. It would also be of interest to consider solutions approaching a time-periodic limit or travelling waves. On a formal level such solutions have been discussed by Cohen, Hoppensteadt and Miura [4].

Remark added in proof :

The author has meanwhile obtained some results in that direction. An extension of this paper is in preparation.

I. BIFURCATION OF SINGULAR SOLUTIONS IN REVERSIBLE SYSTEMS

1. FORMULATION OF THE PROBLEM

We consider a differential equation

$$\frac{du}{dx} = \dot{u} = A(\mu)u + B(\mu,u) \qquad (1.1.)$$

where μ is a real parameter and u is in a Banach space Y. We assume :

(i) $A(\mu)$ is of the form $A(\mu) = A_o + A_1(\mu)$, where $A_o = A(0)$ is a closed, densely defined linear operator in Y and $A_1(\mu) \in \mathcal{L}(Y)$ is a C^∞-function of μ.

(ii) $B : R \times Y \to Y$ is of class C^∞ and $\|B(\mu,u)\| = O(\|u\|^2)$.

(iii) Equation (1.1.) is reversible in the sense of Moser [20], i.e. there exists a linear isometry $R \in \mathcal{L}(Y)$ such that $R^2 = id$, $A(\mu)R = -RA(\mu)$ and $B(\mu,Ru) = -RB(\mu,u)$.

(iv) A_o has an isolated algebraically two-fold but geometrically simple eigenvalue 0. Let N denote the generalized nullspace of A_o, and M the complementary subspace of Y which is invariant under A_o. It easily follows from (iii) that M and N are invariant under R. Moreover, it is not difficult to prove that $R|_N$ has the simple eigenvalues $+1$ and -1.

(v) M has a decomposition $M = M^+ + M^-$, where M^+ and M^- are invariant under A_o, $M^- = RM^+$, and $-A_o|_{M^+}$ generates a strongly continuous semigroup of negative type, i.e. for $x \geq 0$ we have $\|e^{-A_o x}\| \leq Ce^{-\gamma x}$ with positive constants C and γ. It is a simple consequence that on M^- we have $\|e^{A_o x}\| \leq Ce^{-\gamma x}$.

We write $u = (v,w,z)$, where v and w denote the components in N

and $z \in M$. Without restricting generality we may assume that R takes (v,w) to $(v,-w)$. Equation (1.1.) is then rewritten as follows :

$$\dot{v} = \alpha(\mu)w + \gamma(\mu)vw + \sigma(\mu)w^3 + wb^*(\mu)z + O(|v|^2|w| + |w|^3|v| + \|z\|(|\mu| + |v| + |w|\,\|z\| + |w|^2))$$

$$\dot{w} = \beta(\mu)v + \delta(\mu)v^2 + \zeta(\mu)w^2 + O(|v|^3 + |w|^2|v| + \|z\|\cdot(|v| + |w| + \|z\| + |\mu|)) \qquad (1.2.)$$

$$\dot{z} = \tilde{A}(\mu)z + w^2a(\mu) + O(\|z\|(|v| + |w| + \|z\|) + |v|^2 + |v||w| + |w|^3 + |\mu|(|v| + |w|))$$

where $\alpha(\mu)$, $\beta(\mu)$, $\delta(\mu)$, $\zeta(\mu)$ and $\sigma(\mu)$ are real numbers, $a(\mu) \in M$, $b^*(\mu) \in M^*$ (the dual of M), and $A(\mu)$ is a linear operator in M.

We shall distinguish between the following generic cases :

Case 1 :

$$\beta_o = \beta(0) = 0, \quad \beta_1 = \frac{d}{d\mu}\beta(\mu)\big|_{\mu=0} \neq 0, \quad \alpha_o = \alpha(0) \neq 0, \quad \delta_o = \delta(0) \neq 0.$$

Then we put $\mu = \pm\varepsilon$ and introduce the scaling

$v \to \varepsilon^2 v$, $w \to \varepsilon^3 w$, $z \to \varepsilon^3 z$, $x \to \frac{x}{\varepsilon}$. We obtain :

$$\dot{v} = \alpha_o w + O(|\varepsilon|)$$

$$\dot{w} = \pm\beta_1 v + \delta_o v^2 + O(|\varepsilon|) \qquad (1.3.)$$

$$\varepsilon\dot{z} = \tilde{A}(0)z + O(|\varepsilon|)$$

Case 2 :

$$\alpha_o = \alpha(0) = 0, \quad \alpha_1 = \frac{d}{d\mu}\alpha(\mu)\big|_{\mu=0} \neq 0, \quad \beta(0) \neq 0, \quad \beta(0)\tilde{\sigma}(0) - \gamma(0)\zeta(0) \neq 0,$$

where $\tilde{\sigma}(0) = \sigma(0) - b^*(0)\tilde{A}(0)^{-1}a(0)$.

We then put $z' = z - w^2\tilde{A}(0)^{-1}a(0)$ and introduce the scaling

$v \to \varepsilon^2 v$, $w \to \varepsilon w$, $z' \to \varepsilon^2 z'$, $x \to \frac{x}{\varepsilon}$. We obtain

$$\dot{v} = \pm\alpha_1 w + \gamma_o vw + \tilde{\sigma}_o w^3 + O(\|z'\| + |\varepsilon|)$$

$$\dot{w} = \beta_o v + \zeta_o w^2 + O(|\varepsilon|)$$
$$\varepsilon\dot{z}' = \widetilde{A}(0)z' + O(|\varepsilon|) \tag{1.4.}$$

2. SINGULAR SOLUTIONS FOR $\varepsilon = 0$

We are now going to discuss the existence of trajectories connecting saddle points for $\varepsilon = 0$. Since $\widetilde{A}(0)$ is nonsingular, $\varepsilon = 0$ immediately yields $z = 0$ both for (1.3.) and (1.4.), and we are left with a two dimensional problem in either case. We start with the easier case 1.

Case 1 :

We assume $\alpha_o > 0$, $\beta_1 < 0$, which can easily be achieved by replacing μ and v by their negative values if necessary. For $\varepsilon = 0$ (1.3.) reads

$$\dot{v} = \alpha_o w$$
$$\dot{w} = \pm\beta_1 v + \delta_o v^2 \tag{2.1.}$$

which is a Hamiltonian system, i.e. the Hamiltonian

$$H = w^2 \mp \frac{\beta_1}{\alpha_o} v^2 - \frac{2\delta_o}{3\alpha_o} v^3$$

is constant along trajectories. This implies the following

PROPOSITION 2.1. : If the plus sign is chosen in (2.1.) (corresponding to $\mu > 0$), then the fixed point $v = -\beta_1/\delta_o$, $w = 0$ is a saddle point, which is connected to itself by a separatrix, if the minus sign is chosen, the same holds for the fixed point 0.

Case 2 :

(1.4.) reads for $\varepsilon = 0$:

$$\dot{v} = \pm\alpha_1 w + \gamma_o vw + \tilde{\sigma}_o w^3$$
$$\dot{w} = \beta_o v + \zeta_o w^2 \tag{2.2.}$$

Again we may assume $\beta_o > 0$, $\alpha_1 < 0$. Stationary solutions of (2.2.) are given by

$$v = w = 0 \quad \text{or} \quad v = \pm\alpha_1\zeta_o/(\beta_o\tilde{\sigma}_o-\gamma_o\zeta_o), \quad w^2 = \mp\alpha_1\beta_o/(\beta_o\tilde{\sigma}_o-\gamma_o\zeta_o)$$

Hence nontrivial fixed points exist for $\mu > 0$ (i.e. for the choice of the + sign in (2.2.)) iff

$$\beta_o\tilde{\sigma}_o-\gamma_o\zeta_o > 0. \tag{2.3.}$$

A simple calculation shows that in this case the nontrivial fixed points are saddle points. Separatrices connecting the two nontrivial fixed points are found as follows :

We look for invariant parabolae of the form $v = aw^2 + b$. This curve is invariant under the flow of the differential equation (2.2.) iff : $\dot{v} = 2aw\dot{w}$.

If (2.2.) is inserted into this equation, a short calculation shows that there exist in fact two invariant parabolae for (2.2.), namely, we get

$$a = \frac{\frac{1}{2}\gamma_o-\zeta_o \pm \sqrt{(\frac{1}{2}\gamma_o-\zeta_o)^2+2\tilde{\sigma}_o\beta_o}}{2\beta_o}$$

$$b = \frac{\pm\alpha_1}{-\frac{1}{2}\gamma_o-\zeta_o \pm \sqrt{(\frac{1}{2}\gamma_o-\zeta_o)^2+2\tilde{\sigma}_o\beta_o}}$$

It can be checked that the nontrivial fixed points are in fact on these parabolae. We leave the calculations to the reader. That the two parabolae are the only trajectories connecting the two saddle points follows from the uniqueness of stable and unstable manifolds, for which we refer to [22] . Altogether we find.

<u>PROPOSITION 2.2.</u> : If (2.3.) is satisfied and the + sign is chosen in (2.2.) (which corresponds to $\mu > 0$), then there exist two saddle points, which are connected to each other by two trajectories.

The solutions of (2.1.) and (2.2.) are illustrated by the following diagrams:

(2.1.), + sign

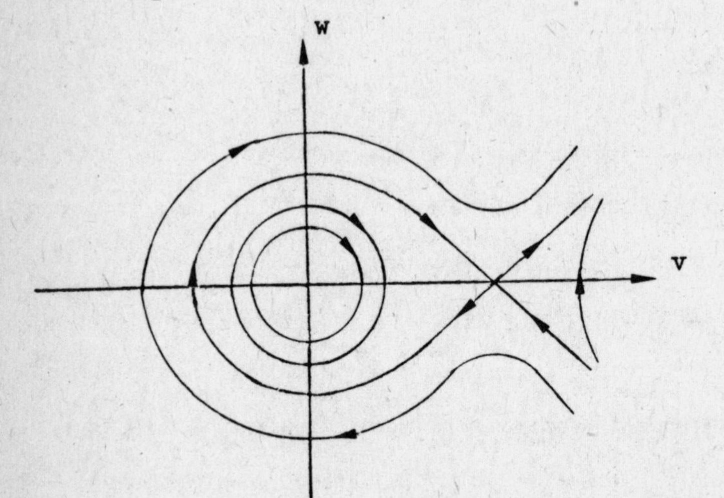

Fig. 2.1

(2.1.), - sign

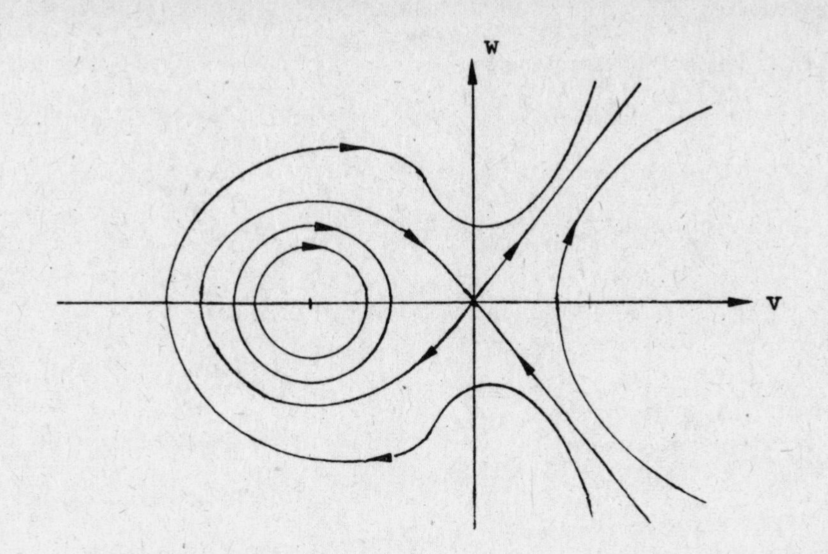

Fig. 2.2.

(2.2.), + sign, (2.3.) holds

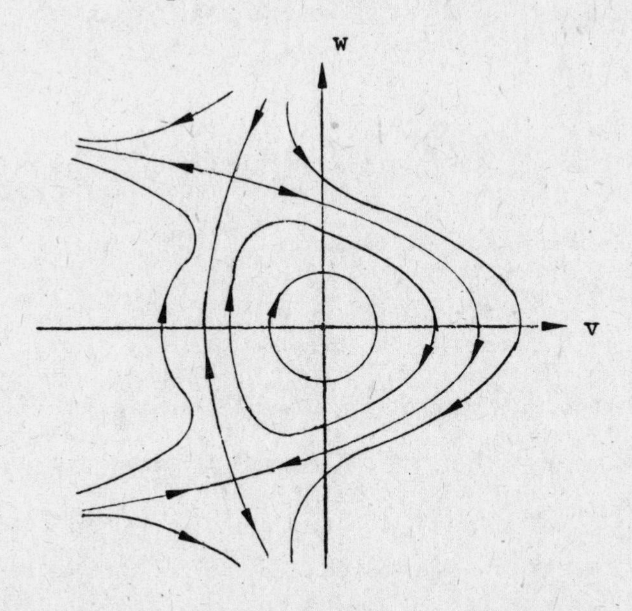

Fig. 2.3.

3. THE LINEARIZATION AT THE SINGULAR SOLUTIONS

All the trajectories connecting saddle points which we have found in §2 are symmetric with respect to the w-axis. This means that among the oneparameter family of solutions represented by such a trajectory there is one solution $y_o(x) = (v_o(x), w_o(x))$ satisfying $Ry_o(x) = y_o(-x)$. We denote the linearization of (2.1.) resp. (2.2.) at $y_o(x)$ by

$$\dot{y} = C_o(x)y$$

Using the reversibility and the fact that $Ry_o(x) = y_o(-x)$ one finds : $C_o(-x)R = -RC_o(x)$. We shall prove

<u>THEOREM 3.1.</u> : For each $f(x) = (f_1(x), f_2(x)) \in F_m = \{f \in C_b^m(R,R^2) |\ Rf(x) = -f(x),$ $\lim_{x\to\infty} f^{(k)}(x),\ \lim_{x\to-\infty} f^{(k)}(x)$ exist for $0 \le k \le m\}$ there exists one and only one $y(x) = (v(x), w(x)) \in U_{m+1} = \{y \in C_b^{m+1}(R,R^2) |\ Ry(x) = y(-x),$ $\lim_{x\to\infty} y^{(k)}(x)\ ,\ \lim_{x\to-\infty} y^{(k)}(x)$ exist for $0 \le k \le m+1\}$ solving the inhomogeneous equation

$$\dot{y} - C_o(x)y = f \tag{3.1.}$$

Here $C_b^m(R,R^2)$ denotes the Banach space of all functions $R \to R^2$ having m continuous bounded derivatives.

<u>Proof</u> : For $x \to \pm\infty$, $y_o(x)$ converges to a saddle point, and one easily concludes from the stable manifold theorem that the convergence is exponential. Repeated differentiation of (2.1.) resp. (2.2.) then yields the result that all derivatives of y_o converge to 0 exponentially. This implies the follo-

wing properties of $C_o(x)$: For $x\to\pm\infty$ $C_o(x)$ converges to the linearization at a saddle point, i.e. $\lim_{x\to\infty} C_o(x)$ exists, and this matrix has one positive and one negative eigenvalue. Moreover, $d^m/dx^m\, C_o(x)$ converges to 0 exponentially for each $m > 0$. From this we see that it is sufficient to prove the theorem for $m = 0$, the rest following from (3.1.) by repeated differentiation. We rewrite (3.1.) in the form

$$\dot{y} - \hat{C}y - \tilde{C}_o(x)y = f$$

where $\hat{C}$ is constant, and $\lim_{x\to\infty} \tilde{C}_o(x) = 0$. From the fact that $\hat{C}$ has one positive and one negative eigenvalue we conclude that for each $f \in C_b([X,\infty),R^2)$ the equation

$$\dot{y} - \hat{C}y = f \qquad (3.2.)$$

has a solution $y \in C_b^1([X,\infty),R^2)$, which is determined up to one free initial condition at $x = X$. If $\lim_{x\to\infty} f(x)$ exists, then $\lim_{x\to\infty} y(x)$ exists as well. Let P denote a projection of R^2 onto a one-dimensional subspace such that the prescription of $Py(X)$ determines the solution of (3.2.) uniquely. The mapping $y \to (\dot{y} - \hat{C}y, Py(X))$ is an isomorphism from $U_X = \{y \in C_b^1([X,\infty),R^2) \,|\, \lim_{x\to\infty} y(x),\ \lim_{x\to\infty} \dot{y}(x) \text{ exist}\}$ onto $(F_X = \{f \in C_b([X,\infty)R^2) \,|\, \lim_{x\to\infty} f(x) \text{ exists}\}) \times R$. Since (3.2.) is autonomous, the norm of this isomorphism and its inverse are independent of X. On the other hand, the norm for the mapping $y \to \tilde{C}_o(x)y$ from U_X into F_X tends to 0 as $X\to\infty$, i.e. for sufficiently large X the mapping $y \to (\dot{y}-Cy-\tilde{C}_o(x)y, Py(X))$ is still an isomorphism from U_X onto $F_X \times R$. That means, given any $f \in F_o$, there exists a bounded solution y to (3.1.) on the interval $[X,\infty)$, which is determined up to one initial condition at $x = X$. Since the initial value problem for (3.1.) is uniquely solvable, this solution on $[X,\infty)$ extends to a solution on $[0,\infty)$, and we have one free initial condition at $x = 0$. This remaining free initial condition is matched by adding a multiple of the solution $\dot{y}_o(x)$ of the homogeneous problem. The requirement $y(x) = Ry(-x)$ implies $w(0) = 0$, which is achieved by one and only one choice of the initial condition, since $\dot{w}_o(0) \neq 0$. This determines the "half-sided" solution on $[0,\infty)$ uniquely. Since $Rf(x) = -f(-x)$, we have further : If $y(x)$ solves (3.1.) on $[0,\infty)$, then $Ry(-x)$ solves (3.1.) on $(-\infty,0]$ (use the fact that $C_o(-x)R = -RC_o(x)$). Hence the half-sided solution extends to a solution on all of R, and the theorem is proved.

4. EXISTENCE OF SINGULAR SOLUTIONS FOR $\varepsilon \neq 0$

The goal of this paragraph is to prove that there exists a branch of singular solutions in a neighbourhood of $\varepsilon = 0$. For this let $y_o(x)$ be as in §3 and put $h(x) = (v(x),w(x)) - y_o(x)$. (1.3.) or (1.4.) then takes the form

$$\begin{aligned} \dot{h} &= C_o(x)h + f(\varepsilon,h,z,x) \\ \varepsilon\dot{z} &= \tilde{A}(o)z + g(\varepsilon,h,z,x) \end{aligned} \qquad (4.1.)$$

Here $C_o(x)$ is as in §3 and we have $\|f(\varepsilon,h,z,x)\| = O(|\varepsilon| + \|z\| + \|h\|^2)$, $\|g(\varepsilon,h,z,x)\| = O(|\varepsilon|)$. For each $m \in N$ the mapping $(\varepsilon,h,z) \to (f(\varepsilon,h,z,x), g(\varepsilon,h,z,x))$ is a C^∞-mapping from $R \times U_m(Y)$ to $F_m(Y)$, where we denote

$$U_m(Y) = \{u \in C_b^m(R,Y) \mid \lim_{x\to\infty} u^{(k)}(x),\ \lim_{x\to-\infty} u^{(k)}(x) \text{ exist for } 0 \leq k \leq m,$$
$$Ru(x) = u(-x)\} \quad \text{and}$$

$$F_m(Y) = \{f \in C_b^m(R,Y) \mid \lim_{x\to\infty} f^{(k)}(x), \lim_{x\to-\infty} f^{(k)}(x) \text{ exist for } 0 \leq k \leq m,$$
$$Rf(x) = -f(-x) \ .$$

We write (4.1.) as follows :

$$\begin{aligned} & h - (\frac{d}{dx} - C_o(x))^{-1} f(\varepsilon,h,z,x) = 0 \\ & z - (\varepsilon \frac{d}{dx} - \tilde{A}(0))^{-1} g(\varepsilon,h,z,x) = 0 \end{aligned} \qquad (4.2.)$$

For $\varepsilon = 0$ equation (4.2.) has the solution $h = 0$, $z = 0$. We abbreviate the left hand side of (4.2.) by $G(\varepsilon,h,z)$. The following assertions will be shown to hold :

(i) G is a continuous mapping from a zero neighbourhood in $R \times U_m(Y)$ into $U_m(Y)$.

(ii) $G(0,0,0) = 0$.

(iii) For fixed ε G is continously differentiable with respect to h and z.

(iv) $D_{(h,z)}G(0,0,0) : U_m(Y) \to U_m(Y)$ is an isomorphism.

(v) $D_{(h,z)}G$ is continuous at the point $(0,0,0)$.

Using theorem 5.1., this immediately leads to the following result :

THEOREM 4.1. : For each $m \in N$ there exists a neighbourhood $V^{(m)}$ of $(0,0,0)$ in $R \times U_m(Y)$ such that in $V^{(m)}$ (4.2.) has a unique resolution $h = h(\varepsilon)$, $z = z(\varepsilon)$.

The hypotheses (i) and (iii) are clear, and (iv) follows from §3. To prove (i) and (v) it suffices to show that the following lemma holds :

LEMMA 4.2. : (i) The mapping $(\varepsilon, z) \to (\varepsilon \frac{d}{dx} - \tilde{A}(0))^{-1} z$ from $R \times F_m(M)$ into $U_m(M)$ is continuous (M has been defined in (iv) of §1).

(ii) The operator norm of $(\varepsilon \frac{d}{dx} - \tilde{A}(0))^{-1} : F_m(M) \to U_m(M)$ has a uniform bound in ε.

Proof : Concerning (i) it is enough to consider the case $m = 0$, since

$$\frac{d}{dx} (\varepsilon \frac{d}{dx} - \tilde{A}(0))^{-1} z = (\varepsilon \frac{d}{dx} - \tilde{A}(0))^{-1} \frac{dz}{dx}$$

Moreover, we shall use the decomposition $M = M^+ + M^-$, which naturally induces a decomposition $C_b^{lim}(R,M) = C_b^{lim}(R,M^+) + C_b^{lim}(R,M^-)$.
(Here $C_b^{lim}(R,M)$ denotes the space of all functions in $C_b(R,M)$ having a limit for $x \to \pm\infty$). For $z \in C_b^{lim}(R,M^+)$ the unique bounded solution of

$$\varepsilon \frac{dr}{dx} - \tilde{A}(0) r = z$$

is given by the representation

$$r(x) = \frac{1}{\varepsilon} \int_{\pm\infty}^{x} \exp(\varepsilon^{-1} \tilde{A}(0)(x-s))\, z(s)\, ds \qquad (4.3.)$$

where the sign in the lower bound equals the sign of ε. Hence we find

$$\begin{aligned} \|r(x)\| &\le \frac{1}{\varepsilon} \sup_{s \in R} \|z(s)\| \, C \int_{\pm\infty}^{x} \exp(\frac{\gamma}{\varepsilon}(x-s))\, ds \\ &\le \frac{C}{\gamma} \sup_{s \in R} \|z(s)\| \end{aligned}$$

and hence (ii). Next, since $\lim_{s \to \infty} z(s) =: z_\infty$ exists, we may write

$$r(x) = \frac{1}{\varepsilon} \int_{\pm\infty}^{x} \exp(\varepsilon^{-1} \tilde{A}(0)(x-s)) z_\infty \, ds + \int_{\pm\infty}^{x} \exp(\varepsilon^{-1} \tilde{A}(0)(x-s))(z(s) - z_\infty)\, ds.$$

The first term is equal to $-\tilde{A}(0)^{-1} z_\infty$, the second can be estimated by

$\frac{C}{\gamma} \sup_{s\in[x,\infty)} \|z(s)-z_\infty\|$. This proves $\lim_{x\to-\infty} r(x) = -\tilde{A}(0)^{-1}z_\infty$.

A similar argument proves that $\lim_{x\to-\infty} r(x)$ exists, i.e. that r is actually in $C_b^{\lim}(R,M^+)$. To complete the proof of (i) we have to show the continuity. (The reversibility condition in the definitions of U_m and F_m is of purely algebraic nature and easy to check). This continuity follows from

(ii) and

(i)' for each given $z \in C_b^{\lim}(R,M^+)$ the solution r given by (4.3.) depends continuously on ε.

We first prove continuity at $\varepsilon = 0$. Let $\varepsilon > 0$. From (4.3.) we conclude

$$r(x) = \frac{1}{\varepsilon}\int_\infty^x \exp(\varepsilon^{-1}\tilde{A}(0)(x-s))z(s)ds$$
$$= -\tilde{A}(0)^{-1}z(x) + \frac{1}{\varepsilon}\int_\infty^x \exp(\varepsilon^{-1}\tilde{A}(0)(x-s))(z(s)-z(x))ds$$
$$= -\tilde{A}(0)^{-1}z(x) + \frac{1}{\varepsilon}\int_\infty^{x+\alpha} \exp(\varepsilon^{-1}\tilde{A}(0)(x-s))(z(s)-z(x))ds$$
$$+ \frac{1}{\varepsilon}\int_{x+\alpha}^x \exp(\varepsilon^{-1}\tilde{A}(0)(x-s))(z(s)-z(x))ds$$

The two integrals in this last expression can be estimated as follows :

$$\left\| \frac{1}{\varepsilon}\int_\infty^{x+\alpha} \dots \right\| \le \frac{2}{\varepsilon}\sup_{s\in R}\|z(s)\| \int_\infty^{x+\alpha} C\exp(\frac{\gamma}{\varepsilon}(x-s))ds$$
$$\le \frac{2C}{\gamma}\sup_{s\in R}\|z(s)\|\exp\left(-\frac{\gamma\alpha}{\varepsilon}\right)$$

$$\left\| \frac{1}{\varepsilon}\int_{x+\alpha}^{x} \dots \right\| \le \frac{1}{\varepsilon}\sup_{|s-s'|\le\alpha}\|z(s)-z(s')\|\, C\int_{x+\alpha}^{x} \exp(\frac{\gamma}{\varepsilon}(x-s))ds$$
$$\le \sup_{|s-s'|\le\alpha}\|z(s)-z(s')\|\frac{C}{\gamma}\left(1-\exp\left(-\frac{\gamma\alpha}{\varepsilon}\right)\right)$$

Due to the uniform continuity of z the second term tends to 0 uniformly in $\varepsilon \ge 0$ as $\alpha \to 0$, and for fixed α the first term tends to 0 as $\varepsilon \to 0$. This proves cotinuity for $\varepsilon\to 0+$, and continuity on the other side is proved analogously, replacing the lower bound ∞ in the integrals by $-\infty$. To prove continuity for $\varepsilon \neq 0$, it is obviously sufficient to prove :

For $\varepsilon > 0$ the expression $\int_{\infty}^{x} \exp(\varepsilon^{-1} \tilde{A}(0)(x-s))z(s)\, ds$ depends continuously on ε, the continuity being uniform with respect to x.

For $\delta < 0$ we have

$$\int_{\infty}^{x} \exp((\varepsilon+\delta)^{-1} \tilde{A}(0)(x-s))z(s)ds - \int_{\infty}^{x} \exp(\varepsilon^{-1} \tilde{A}(0)(x-s))z(s)ds$$

$$= \int_{\infty}^{x} \exp(\varepsilon^{-1} \tilde{A}(0)(x-s))(\exp(-\frac{\delta\tilde{A}(0)}{\varepsilon(\varepsilon+\delta)}(x-s))-1)z(s)ds = \int_{\infty}^{x+A} \dots + \int_{x+A}^{x} \dots$$

It is easily seen that the first integral converges to 0 univormly in δ and x for $A \to \infty$. Using the uniform continuity of z and the strong continuity of the semigroup, one can further prove that for fixed A the second integral converges to 0 as $\delta \to 0$ uniformly with respect to x. This proves one-sided continuity, and the continuity on the other side is proved analogously.

Estimates of the same kind and induction with respect to k can further be used to prove :

LEMMA 4.3. : The mapping $(\varepsilon,z) \to (\varepsilon \frac{d}{dx} - \tilde{A}(0))^{-1} z$ from $R \times F_m(M)$ into $U_{m-k}(M)$ is of class C^k.

For the proof one has to use the fact that $\frac{d}{d\varepsilon} (\varepsilon \frac{d}{dx} - \tilde{A}(0))^{-1}$ $= - (\varepsilon \frac{d}{dx} - \tilde{A}(0))^{-2} \frac{d}{dx}$ etc. We leave the details of the proof to the reader.

We can now conclude from theorem 5.2. (put $X = R$, $Y^{(k)} = U_{m-k}(Y)$) :

THEOREM 4.4. : If the solution $(h(\varepsilon),z(\varepsilon))$ existing by theorem 4.1. is considered as an element of $U_{m-k}(Y)$, then in some neighbourhood of $\varepsilon = 0$ it is a C^k-function of ε.

Remarks : 1. In order to carry out the iteration procedure in the proof of the implicit function theorem, equations of type (3.1.) have to be solved. Although the explicit solution of (3.1.) has not been employed in the proofs, it can easily be obtained modulo integrations, since one integral of the homogeneous equation is known.

2. In the case 1 of §1,2 there occur singular solutions approaching 0 for $x \to \pm\infty$. Using the stable manifold theorem it can be shown the convergence is exponential, and the singular solutions are therefore in L^p for each $p \geq 1$. But do they bifurcate in L^p ? To answer this question, we must see how the scaling introduced in §1 affects the L^p-norm. In the generic case considered here u has been scaled by ε^2 and x by ε^{-1}, which gives a factor of $\varepsilon^{2-1/p}$ in the L^p-norm. Therefore we have a bifurcation in the space L^p for each $p \geq 1$. This need no longer be true if degeneracies occur and different scaling factors must be used. Küpper and Riemer [17] have considered the example

$$- \ddot{u} - |u|^r f(u) = \lambda u$$

where $f(0) < 0$ and $r > 1$.

Our method described above applies to this example when the scaling $\lambda \to -\varepsilon^{-2}\lambda$, $u \to \varepsilon^{2/(r-1)} u$, $x \to \varepsilon^{-1} x$ is used. In the L^p-norm this gives a factor of $\varepsilon^{2/(r-1)-1/p}$. This exponent is greater than zero iff $r < 2p+1$. This agrees with the result obtained in [17] for the case $p = 2$.

5. A GENERALIZATION OF THE IMPLICIT FUNCTION THEOREM

We now prove the generalization of the implicit function theorem that we have used in §4.

<u>THEOREM 5.1.</u> : Let X, Y and Z be Banach spaces, U a neighbourhood of $(0,0)$ in $X \times Y$, and $F : U \to Z$ a mapping having the following properties :

(i) $F(0,0) = 0$

(ii) F is continuous

(iii) F is continuously differentiable with respect to y for each fixed x.

(iv) $D_y F(0,0) : Y \to Z$ is an isomorphism.

(v) $D_y F$ is continuous at the point $(0,0)$.

Then the equation $F(x,y) = 0$ has a unique resolution $y = f(x)$ in some neighbourhood of $(0,0)$.

The only difference to the implicit function theorem as stated in the literature is that we do not assume the continuity of $D_y F$ in a neighbourhood of $(0,0)$. However, a close inspection of the proof of the implicit function theorem shows that the continuity of $D_y F$ in a whole neighbourhood in not really needed and only (iii) and (v) are used. So the classical proof of the implicit function theorem applies without any changes as we demonstrate below.

<u>Proof</u> : Let $A := D_y F(0,0)$. Consider the equation $y = y - A^{-1}F(x,y) =: G(x,y)$
We shall prove : There exist $r, \delta > 0$ such that: $\forall\, x \in B_r$ the mapping $G(x,.)$ takes $\overline{B}_\delta$ into itself and is contracting in $\overline{B}_\delta$. Here B_r denotes the open ball in X centered at 0 and having radius r, and $\overline{B}_\delta$ the closed ball in Y centered at 0 and having radius δ . From Banach's fixed

point theorem we can then conclude that for $x \in B_r$ there is a unique $y = f(x) \in \overline{B}_\delta$ such that $F(x,y) = 0$.

Let now $H := AG$. Then we have

$$H(x,y) - H(x,\eta) = \int_0^1 [D_y F(0,0) - D_y F(x,\eta+t(y-\eta))] \, dt \, (y-\eta)$$

(Here (iii) was used). According to (v) r" and δ can be chosen such that the norm of the integrand is smaller than ε_1 for $\|x\| \le r''$, $\|y\|, \|\eta\| \le \delta$. ε_1 is chosen such that $\|\varepsilon_1 A^{-1}\| < \frac{1}{2}$. Consequently we have

$$\|H(x,y) - H(x,\eta)\| \le \varepsilon_1 \|y-\eta\| \qquad \forall\, x \in B_{r''} \quad \forall\, y,\eta \in \overline{B}_\delta$$

whence

$$\|G(x,y) - G(x,\eta)\| \le \varepsilon_1 \|A^{-1}\| \, \|y-\eta\| < \frac{1}{2} \|y-\eta\|$$

Thus G is contracting. Let now $y \in \overline{B}_\delta$. Then we have

$$\|G(x,y) - G(x,0)\| \le \frac{1}{2} \|y\| \Rightarrow \|G(x,y)\| \le \|G(x,0)\| + \frac{\delta}{2}$$

According to (ii) there exists an $r' > 0$ such that $\|G(x,0)\| \le \frac{\delta}{2}$ for $\|x\| \le r'$. So if we choose $r = \min(r', r'')$, we get the statement above and hence the existence of f. It remains to be shown that f is continuous. Let $x, x+h \in B_r$. Then

$$\begin{aligned}
&\|f(x+h) - f(x)\| \le \|G(x+h,f(x+h)) - G(x,f(x))\| \\
&\le \|G(x+h,f(x+h)) - G(x+h,f(x))\| + \|G(x+h,f(x)) - G(x,f(x))\| \\
&\le \frac{1}{2}\|f(x+h) - f(x)\| + \|G(x+h,f(x)) - G(x,f(x))\| \\
&\Rightarrow \quad \|f(x+h) - f(x)\| \le 2\|G(x+h,f(x)) - G(x,f(x))\|
\end{aligned}$$

According to (ii) G is continuous, and therefore the right side tends to 0 as $h \to 0$.

From the classical implicit function theorem it is known that f is sufficiently regular if F is. We must here, however, deal with problems where F is not regular as a function from $X \times Y$ into Z but only as a function into a larger space than Z. This difficulty is overcome by the next theorem.

THEOREM 5.2. : Let $Y^{(k)}$ resp. $Z^{(k)}$ $(k = 0,\dots,N)$ be two hierarchies of Banach spaces such that $Y^{(k)} \subset Y^{(k+1)}$, $Z^{(k)} \subset Z^{(k+1)}$, the imbeddings being continuous. Let X be a finite dimensional Banach space and F a mapping from a neighbourhood U of 0 in $X \times Y^{(N)}$ into $Z^{(N)}$ having the following properties :

(i) $F(U \cap (X \times Y^{(k)})) \subset Z^{(k)}$ $\quad k = 0,1,\dots,N$

(ii) For each fixed $k, F_k := F|_{U \cap (X \times Y^{(k)})}$ satisfies the conditions of theorem 5.1., when it is considered as a mapping from $X \times Y^{(k)}$ into $Z^{(k)}$. For x fixed $F_k(x,.)$ is a smooth (i.e. sufficiently often differentiable) mapping.

(iii) $F : X \times Y^{(k)} \to Z^{(k+m)}$ is of class C^m for each $k = 0,1,\dots,N$ and $m \leq N - k$.

(iv) The mapping $(x,y,u^1,\dots,u^j) \to z = D_{x^i y^j} F(x,y)(u^1,\dots,u^j)$ is a continuous mapping from $X \times Y^{(k)} \times (Y^{(k)})^j$ into $\mathcal{L}^i(X,Y^{(k+i)})$.

Then the following holds :

The solution $y = f(x) \in Y^{(0)}$ existing by theorem 5.1. is a C^m-function of x in some neighbourhood V_m of 0, if y is regarded as an element of $Y^{(m)}$.

Proof : Let us first consider the case $m = 1$. According to (iii) $F : X \times Y^{(0)} \to Z^{(1)}$ is continuously differentiable. We have :

$$F(x+h,y+\ell) - F(x,y) = F(x+h,y+\ell) - F(x,y+\ell) + F(x,y+\ell) - F(x,y)$$

and

$$F(x,y+\ell) - F(x,y) = D_y F(x,y)\,\ell + R_1(\ell)$$

where

$$\lim_{\|\ell\|_1 \to 0} \|R_1(\ell)\|_1 \Big/ \|\ell\|_1 = 0.$$

(The index 1 refers to the norms in $Y^{(1)}$ and $Z^{(1)}$ resp.).

Using the mean value theorem we find further

$$\|F(x+h,y+\ell) - F(x,y+\ell) - D_x F(x,y+\ell)h\|_1$$
$$\leq \max_{0 \leq \lambda \leq 1} \|D_x F(x+\lambda h,y+\ell) - D_x F(x,y+\ell)\|_1 \|h\|$$

and (iii) yields

$$\|D_x F(x,y+\ell) - D_x F(x,y)\|_1 \to 0 \quad \text{as} \quad \|\ell\|_o \to 0$$

Putting things together we find

$$F(x+h,y+\ell) - F(x,y) = D_x F(x,y)h + D_y F(x,y)\ell + R(h,\ell)$$

where

$$\|R(h,\ell)\|_1/(\|h\|+\|\ell\|_1) \to 0 \quad \text{as} \quad \|h\| \to 0, \|\ell\|_o \to 0.$$

If we put especially $\ell = f(x+h) - f(x)$, we get

$$0 = F(x+h,f(x+h)) - F(x,f(x)) = D_x F(x,f(x))h + D_Y F(x,f(x)) + R(h,\ell)$$

and therefore

$$\ell = -D_y F(x,f(x))^{-1} D_x F(x,f(x))h - D_y F(x,f(x))^{-1} R(h,\ell)$$
$$=: -Lh + (D_y F)^{-1} R$$

(In all these formulae D_x only means the explicit differentiation with respect to x and not implicit differentiation of f). Apparently D_xF maps X into $Z^{(1)}$ and $(D_yF)^{-1}$ maps $Z^{(1)}$ into $Y^{(1)}$. We now prove

$$\| (D_yF)^{-1} R \|_1 / \| h \| \to 0 \quad \text{as} \quad \| h \| \to 0.$$

To do this, let c_ρ be such that $\| D_yF(x,f(x))^{-1} \|_1 \leq c_\rho$ for $\| x \| \leq \rho$ (here $\| \cdot \|_1$ denotes the operator norm from $Z^{(1)}$ to $Y^{(1)}$). Put

$$Q := c_\rho \sup_{\| x \| \leq \rho} \| D_xF(x,f(x)) \|_1$$

Let $\varepsilon_1 > 0$ be given and choose s such that

$$\| h \| \leq s, \| \ell \|_0 \leq s \Rightarrow \| R(h,\ell) \|_1 \leq \varepsilon_1 (\| h \| + \| \ell \|_1)$$

Then : $\| L \| \leq Q \quad \forall x \in \overline{B}_\rho$ and $\| \ell \|_1 \leq Q \| h \| + \varepsilon_1 c_\rho (\| h \| + \| \ell \|_1)$.

If ε_1 is small enough, this yields the existence of a γ such that $\| \ell \|_1 \leq \gamma \| h \|$. Consequently $\| R(h,\ell) \|_1 \leq \varepsilon_1 (1+\gamma) \| h \|$.

Since ε_1 can be chosen arbitrarily small, we find

$$\lim_{\| h \| \to 0} \| R(h,\ell) \|_1 / \| h \| = 0.$$

Hence $f'(x) = -L \in \mathcal{L}(X, Y^{(1)})$ exists and is given by

$$f'(x) = -D_yF(x,f(x))^{-1} D_xF(x,f(x))$$

$D_xF(x,f(x)) \in \mathcal{L}(X, Z^{(1)})$ depends continuously on x and has finite dimensional range. One further concludes from the continuity of the mapping $(x,u) \to D_yF(x,f(x))u$ that also the mapping $(x,v) \to D_yF(x,f(x))^{-1}v$ is continuous. This yields the continuity of f'. Hence everything is proved for $m = 1$.

For $m \geq 2$ we proceed by induction. Formal differentiation of the equation $F(x,f(x)) = 0$ with respect to x yields an equation having the form

$$D_y F(x,f(x)) f^{(m)}(x) + \sum D_x{}^i{}_y{}^j F(x,f(x))(f^{(n_1)},\ldots,f^{(n_j)}) = 0$$

where

$$\sum_{\alpha=1}^{j} n_\alpha = m-i \quad \text{and} \quad n_\alpha < m.$$

So formally we find the moth derivative

$$f^{(m)}(x) = h(x) := -(D_y F(x,f(x))^{-1} \sum \ldots$$

The theorem now follow from the next two statements :

α) h is a continuous function from X into $\mathscr{L}^{(m)}(X,Y^{(m)})$.

β) h is the derivative of $f^{(m-1)}$ with respect to x.

Ad α) : Since $D_y F(x,f(x))^{-1} u \in Y^{(m)}$ depends continuously on $x \in X$ and $u \in Z^{(m)}$, it is sufficient to prove that all the terms under the sum are continuous functions from X into $\mathscr{L}^{(m)}(X,Z^{(m)})$. By the induction assumption $f^{(\alpha)}$ is continuous from X to $Y^{(\alpha)}$ and a fortiori from X to $Y^{(m-i)}$. The rest follows from (iv).

Ad β) : We have

$$f^{(m-1)}(x) = D_y F(x,f(x))^{-1} \sum \ldots =: D_y F(x,f(x))^{-1} H(x)$$

$H(x)$ is a differentiable mapping from X into $\mathscr{L}^{m-1}(X,Z^{(m)})$. Furthermore $(D_y F(x,f(x))^{-1}$ is differentiable as a function from X into $\mathscr{L}(Z^{(m)},Y^{(m+2)})$

This derivative must coincide with the formal derivative.

II.

6. BIFURCATION OF TRANSIENT SOLUTIONS

We now consider a non-reversible differential equation

$$\frac{du}{dx} = \dot{u} = A(\mu)u + B(\mu,u) \qquad (6.1.)$$

where again μ is a real parameter and u is in a Banach space Y. We assume :

(i) $A(\mu)$ has the form $A(\mu) = A_o + A_1(\mu)$, where $A_o = A(0)$ is a closed densily defined linear operator in Y and $A_1(\mu) \in \mathcal{L}(Y)$ is a C^∞-function of μ.

(ii) $B : R \times Y \to Y$ is of class C^∞ and satisfies $\|B(\mu,u\| = O(\|u\|^2)$.

(iii) A_o has an isolated simple eigenvalue 0. By N we denote the null-space and by M the complementary subspace of Y which is invariant under A_o, and $-A_{o|M^+}$ and $A_{o|M^-}$ generate strongly continuous semi-groups of negative type.

We now write $u = (v,z)$, where v denotes the component in N and z the component in M. (6.1.) has now the form

$$\begin{aligned} \dot{v} &= \alpha(\mu)v + \beta(\mu)v^2 + O(|v|^3 + |v|\,\|z\| + \|z\|^2 + |\mu|\,\|z\|) \\ & \qquad\qquad\qquad\qquad\qquad\qquad\qquad\qquad\qquad (6.2.) \\ \dot{z} &= \tilde{A}(\mu)z + va(\mu) + v^2c(\mu) + O(|v|\,\|z\| + \|z\|^2 + |v|^3) \end{aligned}$$

We have $\alpha(0) = 0$, and we shall assume that $\frac{d}{d\mu}\alpha(\mu)_{|\mu=0} \neq 0$, $\beta(0) \neq 0$. Then we put $z' = z - \tilde{A}(0)^{-1}(va(\mu) + v^2c(\mu))$ and introduce the scaling $v \to \mu v$, $z \to \mu^2 z$, $x \to \mu^{-1}x$. This yields an equation of the form

$$\dot{v} = \alpha_1 v + \beta_o v^2 + O(|\mu|)$$

$$\mu\dot{z}' = \tilde{A}(0)z' + O(|\mu|) \tag{6.3.}$$

where $\beta_o = \beta(0)$, $\alpha_1 = \frac{d}{d\mu}\alpha(\mu)\,|_{\mu=0}$. For $\mu = 0$ equation (6.3.) has the stationary points $z = 0$, $v = 0$ and $z = 0$, $v = -\alpha_1/\beta_o$. Moreover, there exists a bounded trajectory $v = v_o(x)$, $z = 0$ joining these two points. We now consider the linearization at $(v_o(x),0)$, i.e. we investigate the inhomogeneous equation

$$\dot{v} - \alpha_1 v - 2\beta_o v v_o(x) = f(x)$$

$$-\tilde{A}(0)z = g(x) \tag{6.4.}$$

The following holds

<u>LEMMA 6.1.</u> : For each $(f(x),g(x)) \in C_b^{lim}(R,Y)$ equation (6.4.) has a one-parameter family of solutions $(v(x),z(x))$.

<u>Proof</u> : The second equation of (6.4.) yields $z(x) = -\tilde{A}(0)^{-1}g(x)$, and so we are only concerned with the first equation. Since $v_o(x)$ approaches an attracting fixed point for $x \to \infty$ and a repelling one for $x \to -\infty$, all solutions of this equation are bounded. Since we have a free initial condition, this proves the lemma.

The same analysis as in the previous chapters, which we are not going to repeat, now yields the following theorem :

THEOREM 6.2. : For each μ in some neighbourhood of 0 there exists a one parameter family of non-constant bounded solutions to (6.3.). These solutions approach a fixed point as $x \to \pm\infty$.

Since (6.3.) is autonomous, a solution is only determined modulo a shift of the independent variable x, and the one-parameter family of solutions corresponds in fact to one trajectory. Since the solutions existing by theorem 6.2. connect the fixed point 0 to the bifurcating fixed point, we have called them "transient".

III. MODELS FOR CHEMICAL REACTIONS

7. TIME-INDEPENDENT SINGULAR SOLUTIONS. IMPERFECT BIFURCATION

We consider the following reaction scheme

$$\begin{aligned} A + X &\to 2X \\ B + X &\to Y \\ 2X &\to C \\ 3X + D &\to 4X \\ 2Y + E &\to X \end{aligned} \qquad (7.1.)$$

Where the concentrations of B and E are considered as given, those of A and D serve as control parameters, and those of X and Y are unknown. We treat the problem in a one-dimensional infinite geometry. If we denote the concentrations of X and Y by u_1 and u_2, the reaction (7.1.) is described by the following equations :

$$\frac{\partial u_1}{\partial t} = (\mu-b)u_1 - cu_1^2 + \alpha u_1^3 + eu_2^2 + D_1\ddot{u}_1$$
$$(7.2.)$$
$$\frac{\partial u_2}{\partial t} = bu_1 - eu_2^2 + D_2\ddot{u}_2$$

In order to conform in notation with the previous chapters, a dot will here denote the derivative with respect to space and not to time. We investigate time-independent solutions : $\partial u_1/\partial t = \partial u_2/\partial t = 0$. The equations (7.2.) are rewritten as a first order system :

$$\begin{aligned} \dot{u}_1 &= v_1 \\ \dot{v}_1 &= D_1^{-1}\,((b-\mu)u_1 + cu_1^2 - \alpha u_1^3 - eu_2^2) \\ \dot{u}_2 &= v_2 \\ \dot{v}_2 &= D_2^{-1}\,(eu_2^2 - bu_1) \end{aligned} \qquad (7.3.)$$

where b, c, e, D_1 and D_2 are fixed constants, μ and α are variable parameters. Clearly (7.3.) is reversible under the mapping $R : (u_1, v_1, u_2, v_2)$ $(u_1, -v_1, u_2, -v_2)$. We shall assume for the moment that α and μ are controlled in such a way that always $\alpha+\mu = c$. Then for all values of μ (7.3.) has the constant solution $u_1 = 1$, $u_2 = \sqrt{b/e}$. The matrix $A(\mu)$ of the linearization at this solution has the characteristic equation

$$\begin{vmatrix} \frac{1}{D_1}(b+2\mu-c) - \lambda^2 & -\frac{2}{D_1}\sqrt{e.b} \\ -\frac{b}{D_2} & \frac{2}{D_2}\sqrt{e.b} - \lambda^2 \end{vmatrix} = 0$$

$$\lambda^4 - \lambda^2\left(\frac{2}{D_2}\sqrt{e.b} + \frac{1}{D_1}(b+2\mu-c)\right) + \frac{2}{D_1D_2}\sqrt{e.b}\,(2\mu-c) = 0$$

An algebraically two-fold but geometrically simple eigenvalue 0 occurs at $\mu = \frac{c}{2}$. The derivative of λ^2 with respect to μ is different from zero, and the two remaining eigenvalues of $A(\frac{c}{2})$ are non-imaginary. It can further be verified that the condition corresponding to $\delta_o \neq 0$ in §1 is satisfied. (That means that the quadratic nonlinearaty is not in the range of $A(\frac{c}{2})$ when the eigenvector belonging to $\lambda = 0$ is used in the arguments). Thus all the hypotheses of §1, case 1 are satisfied. Therefore theorems 4.1. and 4.2. lead to.

THEOREM 7.1. : On either side of $\mu = \frac{c}{2}$ there exists a branch of singular solutions bifurcating from the point $u_1 = 1$, $u_2 = \sqrt{b/e}$. These singular solutions depend smoothly on the parameter $\varepsilon = \sqrt{\pm(\mu - c/2)}$.

Finally we wish to get rid of the unnatural hypothesis that $\alpha + \mu = c$. For that purpose we put $\delta = \mu - c + \alpha$. If in addition to the scaling introduced in §1 we replace δ by $\varepsilon^4\delta$, the analysis of §§ 2-5 still applies and shows the existence of a singular solution depending on ε and δ in some neighbourhood of $\varepsilon = \delta = 0$. Since $\mu - \frac{c}{2}$ has been scaled by ε^2 and δ by ε^4, we see that the "imperfection" δ has to be small compared to $\mu - \frac{c}{2}$. Nevertheless we find the existence of an open set of parameter values for which singular solutions exist, so they may be of physical significance.
The term "imperfect" bifurcation is understandable if we look at the behaviour of stationary solutions for $\delta \neq 0$ (cf. [24]). For $\delta = 0$ we have the fixed point $u_1 = 1$, from which a second constant solution bifurcates at $\mu = \frac{c}{2}$:

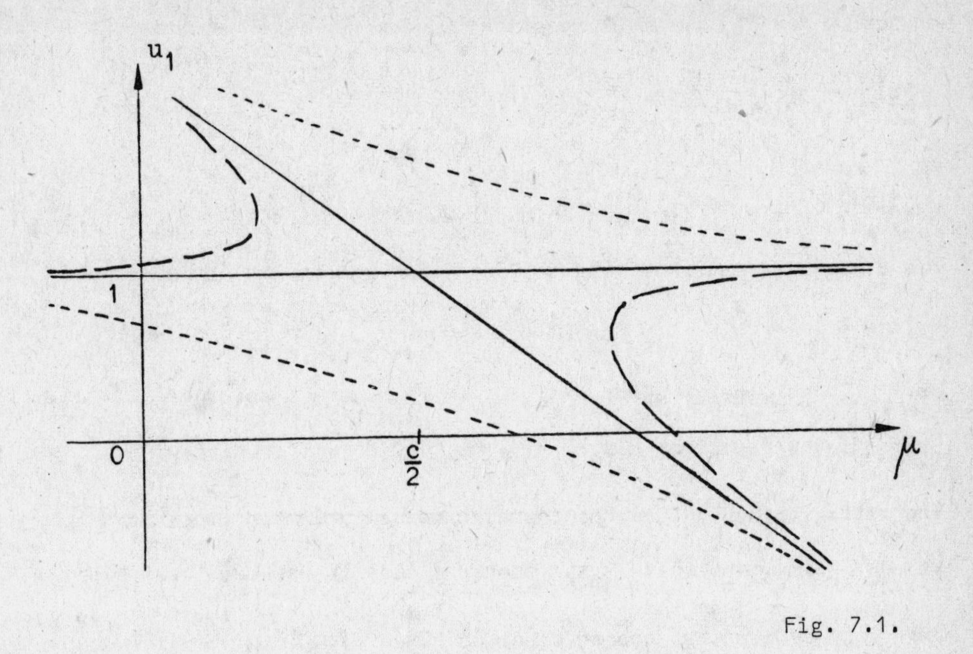

Fig. 7.1.

For $\delta > 0$ one finds fixed points as indicated by the dashed line, and for $\delta < 0$ as indicated by the dotted line. I.e. no bifurcation is present any longer if $\delta \neq 0$. Nevertheless singular solutions still persist.

Remark : In some models for chemical reactions there is a trivial solution for which all concentrations are zero, and one would therefore study singular solutions bifurcating from zero. Only those solutions, however, which give positive concentrations, are acceptable. It would be interesting whether criteria can be found that bifurcating singular solutions satisfy this physical restriction.

8. OSCILLATING SINGULAR SOLUTIONS CONNECTED WITH HOPF BIFURCATIONS

We consider a general chemical reaction model given by an equation

$$\frac{\partial u}{\partial t} = F(\mu,u) + D\,\frac{\partial^2 u}{\partial x^2}$$

where $u \in R^n$, $\mu \in R$. F is some nonlinear function of polynomial character, and D is a diagonal matrix having only positive entries in its diagonal. We shall assume :

(i) For μ in some neighbourhood of 0 there exists a solution $u = u_o(\mu) \in R^n$ of the equation $F(\mu,u) = 0$. u_o is a C^∞-function of μ.

(ii) The matrix $D_u F(0,u_o(0))$ has the algebraically simple imaginary eigenvalues $\pm i\omega_o$, the rest of the spectrum lies in the left half plane.

(iii) Let $\lambda(\mu)$ denote that branch of eigenvalues of $D_u F(\mu,u_o(\mu))$ which passes through $i\omega_o$ at $\mu = 0$. Then we have $\frac{d}{d\mu}\,\mathrm{Re}\,\lambda(\mu)\big|_{\mu=0} > 0$.

(iv) For each $\gamma > 0$ the spectrum of $D_u F(0,u_o(0)) - D\gamma$ is contained in the left half plane.

It is well known that under conditions (i)-(iii) a branch of x-independent time-periodic solutions emanates from the point $u = u_o(0)$, $\mu = 0$ [14]. The conditions (i)-(iv) are known to be satisfied in quite a few chemical reaction models, e.g. in the Field-Noyes model of the Belousov-Zhabotinskii reaction [12]. It is further known [13] that in concrete examples the branch of periodic solutions may exist for positive or for negative values of μ. The object of this chapter is the investigation of time-periodic space-dependent solutions of equation (8.1.), which we now write in the form

$$\frac{1}{\omega}\,\ddot{u} = D^{-1}\left(\frac{\partial u}{\partial t} - \frac{1}{\omega}\,F(\mu;u)\right) \qquad (8.2.)$$

Here the factor $\omega > 0$ has been introduced in order to normalize the period to 2π. After appropriate scaling of x we drop $1/\omega$ on the left side and rewrite the equation (8.2.) as a first order system

$$\begin{aligned} \dot{u} &= v \\ \dot{v} &= D^{-1}\left(\frac{\partial u}{\partial t} - \frac{1}{\omega} F(\mu,u)\right) \end{aligned} \tag{8.3.}$$

Let now $\ell^1_\alpha(R^n)$ denote the Banach space of all 2π-periodic functions $y : R \to R^n$ satisfying $\|y\|_\alpha := \sum_{k\in Z} (|k|^\alpha+1)\, \|y^{(k)}\| < \infty$, where the $y^{(k)}$ are the Fourier coefficients of $y : y(t) = \sum_{k\in Z} y^{(k)} e^{ikt}$. We shall treat (8.3.) as an equation of evolution in the Banach space $Y_m = \{(u,v) | u \in \ell^1_m(R^n),$ $v \in \ell^1_{m-1/2}(R^n)\}$, where m is an arbitrary positive integer.
Apparently the mapping $(\mu,\omega,(u,v)) \to (0,-D^{-1}\ \omega^{-1}F(\mu,u))$ is a smooth mapping from $R \times (R \setminus \{0\}) \times Y_m$ into Y_m. Hence all unbounded operators occur only in the linear part. We now discuss the spectrum of the linearization of the right side of (8.3.) at the point $\omega = \omega_o$, $\mu = 0$, $u = u_o(\mu)$.
We are going to prove.

<u>LEMMA 8.1.</u> : Assume (i)-(iv) and (v) stated below hold. Then we have :
The operator $A : (u,v) \to (v, D^{-1}\frac{\partial u}{\partial t} - (D\omega_o)^{-1} D_uF(0,u_o(0))u)$ is a closed operator in Y_m. It has an isolated algebraically four-fold and geometrically two-fold eigenvalue 0. If N denotes the generalized nullspace and M a complementary invariant subspace, then $A_{|M}$ satisfies the bi-semigroup condition (v) of §1.

<u>Proof</u> : We first note that A acts Fourier-componentwise, i.e. we have $A(\sum (u^{(k)},v^{(k)})e^{ikt}) = \sum A^{(k)}(u^{(k)},v^{(k)})e^{ikt}$, where

$$A^{(k)} = \begin{pmatrix} 0 & 1 \\ D^{-1}ik - (D\omega_o)^{-1}D_uF(0,u_o(0)) & 0 \end{pmatrix}$$

Thus the eigenvalues of $A^{(k)}$ are the square roots of the eigenvalues of the matrix

$$D^{-1}ik - (D\omega_o)^{-1}D_uF(0,u_o(0))$$

and $A^{(k)}$ has an imaginary eigenvalue iff this matrix has a negative real eigenvalue. Let now be

$$(D^{-1}ik - (D\omega_o)^{-1}D_uF(0,u_o(0)))y = -\gamma y \qquad \gamma > 0$$

This yields

$$D_uF(0,u_o(0))y - \omega_o\gamma Dy = ik\omega_o y$$

For $\gamma \neq 0$ this is impossible by condition (iv), and $\gamma = 0$ yields $k = \pm 1$, since $\pm i\omega_o$ are the only imaginary eigenvalues of $D_uF(0,u_o(0))$. We see therefore that $A^{(k)}$ has no imaginary eigenvalues for $k \neq \pm 1$, and the only imaginary eigenvalue for $k = \pm 1$ is equal to 0. To prove the statement concerning the multiplicity, it must be shown that $\lambda = 0$ is an algebraically simple eigenvalue of

$$D^{-1}i - (D\omega_o)^{-1}D_uF(0,u(0))$$

It is easy to see that the nullspace of this matrix is spanned by the eigenvector of $D_uF(0,u_o(0))$ to the eigenvalue $i\omega_o$, whence the eigenvalue $\lambda = 0$ is geometrically simple. Assume now that

$$D_uF(0,u_o(0))y = i\omega_o y$$

and

$$y = (D^{-1}i - (D\omega_o)^{-1}D_uF(0,u_o(0)))z$$

This yields

$$\omega_o Dy = (i\omega_o - D_uF(0,u_o(0)))z$$

We shall therefore assume

(v) It y denotes the eigenvector of $D_uF(0,u_o(0))$ to the eigenvalue $i\omega_o$, then Dy is not in the range of $i\omega_o - D_uF(0,u_o(0))$.

Clearly (v) is a generic condition, which guarantees that the eigenvalue is algebraically simple.

We must now verify that the spectrum of A is actually given by the eigenvalues of the $A^{(k)}$ and that (v) of §1 holds. This will be a consequence of the following :

There exists an isomorphism T of Y_m acting Fourier componentwise :

$$T \sum_{k \in Z} (u^{(k)}, v^{(k)})e^{ikt} = \sum_{k \in Z} T^{(k)}(u^{(k)}, v^{(k)})e^{ikt}$$

such that for large $|k|$, let us say for $|k| \geq k_o$ the matrix $(T^{(k)})^{-1}A^{(k)}T^{(k)}$ consists of a diagonal part and a rest term which has a norm of the order of magnitude $|k|^{-1/2}$. Namely put for $|k| \geq k_o$

$$T^{(k)} = \begin{pmatrix} 1 & \sqrt{\frac{D}{ik}} \\ -\sqrt{\frac{ik}{D}} & 1 \end{pmatrix}$$

This yields

$$(T^{(k)})^{-1} = \begin{pmatrix} \frac{1}{2} & -\frac{1}{2}\sqrt{\frac{D}{ik}} \\ \frac{1}{2}\sqrt{\frac{ik}{D}} & \frac{1}{2} \end{pmatrix}$$

$$(T^{(k)})^{-1}A^{(k)}T^{(k)} = \begin{pmatrix} -\sqrt{\frac{ik}{D}} + O(|k|^{-1/2}) & O(|k|^{-1}) \\ O(1) & \sqrt{\frac{ik}{D}} + O(|k|^{-1/2}) \end{pmatrix}$$

i.e. $(T^{(k)})^{-1}A^{(k)}T^{(k)}$ is the sum of a diagonal matrix and a rest term which has an operator norm of the order of magnitude $(|k|^{-1/2})$ (remind the definition of the norm in Y_m).

It is now a simple consequence of the Hille-Yosida theorem [19] that the bi-semigroup property (v) holds for the diagonal part, since the resolvent estimates required are trivial in this case. A perturbation argument shows that the same holds for the full operator $T^{-1}AT$ and hence for A. This concludes the proof of Lemma 8.1.

For simplicity, we assume now that D is a multiple of the identity.

Let again y denote the eigenvector of $D_uF(0,u_o(0))$ to the eigenvalue $i\omega_o$. Then (u,v) can be decomposed in the form

$$(u,v) = (u_o(\mu),0) + \alpha_1(y,0)e^{it} + \bar{\alpha}_1(\bar{y},0)e^{-it} + \alpha_2(0,y)e^{it} + \bar{\alpha}_2(0,\bar{y})e^{-it} + z$$

where $\alpha_1, \alpha_2 \in C$ and z is in M (M was defined in Lemma 8.1.). Equations (8.3.) then in general take the form

$$\dot{\alpha}_1 = \alpha_2$$

$$\dot{\alpha}_2 = \mu a_1\alpha_1 + a_2\alpha_1^2\bar{\alpha}_1 + C_1(z,\alpha_1) + C_2(z,\bar{\alpha}_1) + (\omega^{-1}-\omega_o^{-1})a_3\alpha_1 + \ldots$$

$$\dot{z} = \tilde{A}(0)z + \alpha_1^2 d_1 + \alpha_1\bar{\alpha}_1 d_2 + \bar{\alpha}_1^2 d_3 + \ldots$$

Here a_1 and a_2 are in general complex numbers, whereas a_3 is always purely imaginary. $C_{1,2} : M \times C \to C$ are bilinear operators, and the d_i are vectors in M. $\tilde{A}(0)$ denotes $A_{|M}$. The dots indicate higher order terms. Analogously as in §1 we put $z = z' - \tilde{A}(0)^{-1} \{\alpha_1^2 d_1 + \alpha_1\bar{\alpha}_1 d_2 + \alpha_1^2 d_3\}$ and introduce the scaling $\alpha_1 \to \varepsilon\alpha_1$, $\alpha_2 \to \varepsilon^2\alpha_2$, $\mu = \pm\varepsilon^2$, $z' \to \varepsilon^2 z'$, $\omega^{-1} - \omega_0^{-1} = \varepsilon^2 \hat{\omega}$, $x \to \varepsilon^{-1}x$. We then obtain the equation

$$\begin{aligned} \dot{\alpha}_1 &= \alpha_2 \\ \dot{\alpha}_2 &= \pm a_1\bar{\alpha}_1 + a_2'\alpha_1^2\alpha_1 + a_3\hat{\omega}\alpha_1 + O(|\varepsilon| + \|z'\|) \\ \varepsilon\dot{z}' &= \tilde{A}(0)z' + O(|\varepsilon|) \end{aligned} \tag{8.4.}$$

For $\varepsilon = 0$ this reduces to

$$\ddot{\alpha}_1 = \pm a_1\alpha_1 + a_3\hat{\omega}\alpha_1 + a_2'\alpha_1^2\bar{\alpha}_1 \tag{8.5.}$$

The coefficients a_1, a_2' and a_3 are the same which determine the Hopf bifurcation at order ε^3. It follows from our assumption (iii) that $\operatorname{Re} a_1$ is negative, a_1 and a_2' are in general complex, and a_3 is purely imaginary. $\hat{\omega}$ is an unknown variable which has to be determined.
We now try to solve (8.5.) by the ansatz : $\alpha_1 = re^{i\phi}$, $r = C \operatorname{sech} kx$, $\dot{\phi} = B \tanh(kx)$. After some elementary calculations this leads to

$$\begin{aligned} k^2 - B^2 &= \pm \operatorname{Re} a_1 \\ -2k^2 + B^2 &= C^2 \operatorname{Re} a_2' \\ -2Bk &= \pm \operatorname{Im} a_1 + \hat{\omega} \operatorname{Im} a_3 \\ 3Bk &= C^2 \operatorname{Im} a_2' \end{aligned} \tag{8.6.}$$

From the fourth and second equation of (8.6.) we find that

$$3Bk \operatorname{Re} a_2' + (2k^2 - B^2) \operatorname{Im} a_2' = 0$$

which is solved by $B = \lambda k$, where

$$\lambda = \frac{-3 \operatorname{Re} a_2' \pm \sqrt{9 (\operatorname{Re} a_2')^2 + 8 (\operatorname{Im} a_2')^2}}{-2 \operatorname{Im} a_2'}$$

According to the fourth equation of (8.6.) λ must have the same sign as $\operatorname{Im} a_2'$, which is achieved by choosing the minus sign in the numerator. We now insert $B = \lambda k$ into the first equation of (8.6.) and obtain

$$k^2 = \pm \operatorname{Re} a_1/(1-\lambda^2)$$

We now assume

(vi) $\lambda \neq \pm 1$

Then (8.6.) is solvable by appropriate choice of the sign + or - . Thus for $\varepsilon = 0$ we have found a solution $\alpha_1^o(x)$, which is an even function of x and satisfies $\lim_{x\to\pm\infty} \alpha_1^o(x) = 0$. We have here assumed that $\operatorname{Im} a_2' \neq 0$. If $\operatorname{Im} a_2' = 0$, however, (8.6.) can be solved by putting $B = 0$, provided that $\operatorname{Re} a_2'$ is negative. Therefore we assume

(vii) If $\operatorname{Im} a_2' = 0$, then $\operatorname{Re} a_2'$ is negative.

Now we discuss the linearization of (8.4.) at the solution $\alpha_1^o(x)$, $\alpha_2^o(x) = \dot{\alpha}_1^o(x)$, i.e. as in §3 we consider the inhomogeneous problem

$$\begin{aligned} &\dot{\beta}_1 - \beta_2 = f_1 \\ &\dot{\beta}_2 \mp a_1\beta_1 - a_2'(\alpha_1^o)^2\bar{\beta}_1 - 2a_2'\alpha_1^o\,\bar{\alpha}_1^o\beta_1 - \hat{\omega}a_3\beta_1 = f_2 \end{aligned} \qquad (8.7.)$$

As in §3 the linearized operator is viewed as an operator from

$$U_{m+1} = \{(\beta_1,\beta_2) \in C_b^{m+1}(R,C^2) \mid \lim_{x\to\infty} \beta_i^{(k)}(x), \ \lim_{x\to-\infty} \beta_i^{(k)}(x)$$

$$\text{exist for } 0 \le k \le m+1, \quad \beta_1(x) = \beta_1(-x), \quad \beta_2(x) = -\beta_2(-x)\}$$

into

$$F_m = \{(f_1,f_2) \in C_b^m(R,C^2) \mid \lim_{x\to\infty} \ldots \text{ exist}, \ f_1(x) = -f_1(-x),$$
$$f_2(x) = f_2(-x)\}.$$

The fixed point $0 \in C^2$ is a saddle point having two stable and two unstable directions. Hence the same arguments as in §3 yield the result that for each $f \in F_m$ we have a bounded solution β of (8.7.) on the interval $[0,\infty)$, and two initial conditions at $x = 0$ are free. These initial conditions can be matched by adding multiples of the solutions of the homogeneous problem, which are given by $(\dot{\alpha}_1^0(x), \dot{\alpha}_2^0(x))$ and $(i\alpha_1^0(x), i\alpha_2^0(x))$. In order to extend our solution to the whole real axis, we need as in §3 that $\beta_2(0)$ is zero. Addition of a real multiple of $\dot{\alpha}_2^0(x)$ now allows to make $\operatorname{Re}\beta_2(0)$ equal to zero, but since $i\alpha_2^0(x)$ is zero for $x = 0$, we cannot achieve that $\operatorname{Im}\beta_2(0)$ equals zero. That means, since $\operatorname{Im}\beta_2(0)$ is a linear functional of f, that the linearized operator has now codimension 1.
We have, however, one more unknown variable, namely $\hat{\omega}$. If we linearize with respect to α and $\hat{\omega}$, then we have to consider the inhomogeneous problem

$$\begin{aligned} &\dot{\beta}_1 - \beta_2 = f_1 \\ &\dot{\beta}_2 \mp a_1\beta_1 - a_2'(\alpha_1^0)^2\bar{\beta}_1 - 2a_2'\alpha_1^0\bar{\alpha}_1^0\beta_1 - \hat{\omega}a_3\beta_1 = f_2 - a_3\Omega\alpha_1^0 \end{aligned} \qquad (8.8.)$$

In order to obtain surjectivity, we must now prove that for each given $f \in F_m$

there exists $\Omega \in R$ such that (8.8) has a solution in U_m, i.e. that $(0,-a_3\alpha_1^o)$ is not in the range of the left side of (8.8.). This can be expected to be true under general conditions, in particular, if a_2^1 is real, It can be shown by explicit calculation, however, that the functional

$$(f_1,f_2) \to \int_{-\infty}^{\infty} -\alpha_2^o(x) \text{ Im } f_1(x) + \alpha_1^o(x) \text{ Im } f_2(x)\, dx$$

annihilates the left side of (8.8.), i.e. the range of the operator on the left of (8.8.) is the nullspace of this functional. It is easily seen that $(0,-a_3\, \alpha_1^o(x))$ is not in that nullspace.

We thus see that the linearization of (8.4.) with respect to α, z and ω is surjective. The dimension of the kernel is equal to 1. As before the generalized implicit function theorem of §5 is now used to prove:

<u>THEOREM 8.2.</u> : If the transversality condition mentioned above holds, then for each ε in a neighbourhood of 0 there exists $\omega \in R$ for which (8.4.) has a one-parameter family of non-constant solutions approaching 0 as $x \to \pm\infty$ and even in x.

The solutions in this one-parameter family are only distinguished by a shift of the time variable t.

<u>Remark</u> : For $\varepsilon = 0$ we found a solution having the form

$$u(x,t) = C \text{ sech } (kx) \exp \left(iB \int_0^x \tanh (kx')dx'\right)e^{it}\, y + \text{c.c.}$$

Assume $k > 0$. For $x \to \infty$ the asymptotic form of the solution is given by

$$u(x,t) \sim 2C\, e^{-kx}\, e^{i(t+Bx)}\, y + \text{c.c.}$$

and for $x \to -\infty$ by

$$u(x,t) \sim 2C\, e^{+kx} e^{i(t-Bx)}\, y + \text{c.c.}$$

That means that asymptotically we have damped waves, which propagate in opposite directions for positive and negative x.

Equation (8.5.) could also be solved by the ansatz $r = C \tanh kx$, $\dot{\phi} = B \tanh kx$. It would be interesting to know what happens to these solutions when $\varepsilon \neq 0$. Moreover, it would be of interest to consider (8.5.) in higher dimensions, where $\ddot{\alpha}_1$ has to be replaced by $\Delta\alpha_1$. This would lead to a problem similar to those discussed by Berestycki, Lions and Peletier (see e.g. [29]). Rotationally symmetric solutions in more than one dimension might correspond to the observed targer patterns in the Belousov-Zhabotinskii reaction.

9. NONPERIODIC TRAVELLING WAVES IN CHEMICAL REACTIONS

We shall again consider the model of §7 and specialize to solutions depending only on $x' = x - \gamma t$. For such solutions (7.2.) reads

$$\begin{aligned} \ddot{u}_1 &= D_1^{-1} (- \gamma\dot{u}_1 - (\mu-b)u_1 + cu_1^2 - \alpha u_1^3 - eu_2^2) \\ \ddot{u}_2 &= D_2^{-1} (- \gamma\dot{u}_2 - bu_1 + eu_2^2) \end{aligned} \qquad (9.1.)$$

where the dot now denotes differentiation whith respect to x'. (9.1.) can of course again be written as a first order systems, when we put $\dot{u}_1 = v_1$, $\dot{u}_2 = v_2$. Again we assume that α is controlled in such a way that $\alpha+\mu = c$. Then (9.1.) still has the stationary solution $u_1 = 1$, $u_2 = \sqrt{b/e}$; and the linearization $A(\mu)$ at this solution now yields the characteristic equation

$$\begin{vmatrix} D_1^{-1}(b+2\mu-c) - \lambda^2 - D_1^{-1}\gamma\lambda & -2D_1^{-1}\sqrt{e.b} \\ -b/D_2 & 2D_2^{-1}\sqrt{e.b} - \lambda^2 - D_2^{-1}\gamma\lambda \end{vmatrix} = 0$$

$$\lambda^4 + \lambda^3\gamma(D_1^{-1}+D_2^{-1}) - \lambda^2(2/D_2\sqrt{e.b}+(b+2\mu-c)/D_1 - \gamma^2/(D_1D_2))$$

$$-\lambda \quad 2\gamma/(D_1D_2)\sqrt{e.b} - \gamma\lambda/(D_1D_2)(b+2\mu-c) + 2/(D_1D_2)\sqrt{e.b}\,(2\mu-c) = 0$$

This shows that we now have an algebraically simple eigenvalue $\lambda = 0$ if $\mu = c/2$. It $\gamma \neq 0$, the derivative of λ with respect to μ at this point does not vanish, and it is also easily seen that the remaining eigenvalues of $A(c/2)$ are not imaginary. One can also show by explicit calculation that the quadratic term corresponding to $\beta_o v^2$ in equation (6.3.) does not vanish. We can therefore apply theorem 6.2., which gives the result

THEOREM 9.1. : Let γ be any fixed real number not equal 0. If $\alpha+\mu = c$, then in some neighbourhood of $\mu = c/2$ there exists a branch of transient solutions to (9.1.), i.e. solutions approaching one limit as $x'\to\infty$ and another limit as $x' \to -\infty$.

As in §7, these transient solution persist under an imperfection when $\delta = \alpha+\mu-c$ is scaled by the factor $(\mu-c/2)^2$.

Acknowledgements : I thank Professor K. Kirchgässner and Dr. J. Scheurle for their cooperation, and I thank the Volkswagen Foundation for financial support of this work within the project of Synergetics.

Bibliography

1 J.F.G. Auchmuty and G. Nicolis, Bifurcation Analysis of Reaction-Diffusion Equations III. Chemical Oscillations, Bull. Math. Biol. 38 (1976), pp. 325-350.

2 G.I. Barenblatt and Ya.B. Zeldovich, Intermediate Asymptotics in Mathematical Physics, Russ. Math. Surveys 26,2 (1971), pp. 45-61.

3 R.K. Bullough, Solitons, in : Interaction of Radiation with Condensed Matter, Vol. I, International Atomic Energy Agency, Vienna 1977, pp. 381-469.

4 D.S. Cohen, F.C. Hoppensteadt and R.M. Miura, Slowly Modulated Oscillations in Nonlinear Diffusion Processes, SIAM J. Appl. Math. 33 (1977), pp. 217-229.

5 M.G. Crandall and P.H. Rabinowitz, Bifurcation, Perturbation of Simple Eigenvalues, and Linearized Stability, Arch. Rat. Mech. Anal. 52 (1973), pp. 161-180.

6 R.J. Field and R.M. Noyes, Oscillation in Chemical Systems IV. Limit Cycle Behaviour in a Model of a Real Chemical Reaction, J. Chem. Phys. 60 (1974), pp. 1877-1884.

7 P.C. Fife, Asymptotic Analysis of Reaction-Diffusion Wave Fronts, Rocky Mt. J. Math. 7 (1977), pp. 389-415.

8 P.C. Fife, Asymptotic States for Equations of Reaction and Diffusion, Bull. AMS 84 (1978). pp. 693-726.

9 H. Haken, Synergetics. An Introduction, Springer 1977.

10 S.P. Hastings and J.D. Murray, The existence of Oscillatory Solutions in the Field-Noyes Model of the Belousov-Zhabotinskii Reaction, SIAM J. Appl. Math. 28 (1975), pp. 678-688.

11 P.C. Hemmer and M.G. Velarde, Multiple Steady States for the Degn-Harrison Reaction Scheme of a Bacterial Culture, Z. Physik B. 31 (1978), pp; 111-116.

12 In-Ding Hsü, Existence of Periodic Solutions for the Belousov-Zaikin-Zhabotinskii Reaction, J. Diff. Eq. 20 (1976), pp. 399-403.

13 In-Ding Hsü and N.D. Kazarinoff, An Applicable Hopf Bifurcation Formula and Instability of Small Periodic Solutions of the Field-Noyes Model, J. Math. Anal. Appl. 55 (1976), pp. 61-89.

14 E. Hopf, Abzweigung einer periodischen Lösung eines Differentialsystems, Ber. Sächs. Akad. Wiss., Math.-nat. Kl. 94 (1942).

15 K. Kirchgässner and J. Scheurle, One the Bounded Solutions of a Semilinear Elliptic Equation in a Strip, J. Diff. Eq. 32 (1979), pp. 119-148.

16 K. Kirchgässner and J. Scheurle, Bifurcation from the Continuous Spectrum and Singular Solutions, preprint.

17 T. Küpper and D. Riemer, Necessary and Sufficient Conditions for Bifurcation from the Continuous Spectrum, Nonlinear Analysis 3 (1979), pp. 555-561.

18 N. Kopell and L.N. Howard, Bifurcations and Trajectories Joining Critical Points, Adv. Math. 18 (1975), pp. 306-358.

19 G.E. Ladas and V. Lakshmikantham, Differential Equations in Abstract Spaces, Academic Press 1972.

20 J. Moser, Stable and Random Motions in Dynamical Systems, Princeton University Press 1973.

21 J.D. Murray, On a Model for the Temporal Oscillations in the Belousov-Zhabotinskii Reaction, J. Chem. Phys. 61 (1975), pp. 3610-3613.

22 O. Perron, Uber Stabilität und asymptotisches Verhalten der Integrale von Differentialgleichungs systemen, Math. Z. 29 (1929), pp. 129-160.

23 A.B. Poore, A model Equation Aristing from Chemical Reactor Theory, Arch. Rat. Mech. Anal. 52 (1973), pp. 358-388.

24 M. Potier-Ferry, Perturbed Bifurcation Theory, J. Diff. Eq. 33. (1979), pp. 112-146.

25 M. Renardy, Quasiperiodische Verzweigungslösungen rotationssymmetrischer Probleme, Dissertation Universität Stuttgart 1980.

26 M. Renardy On Bounded Solutions of a Classical Yang-Mills Equation, to appear in Comm. Math. Phys.

27 M.M. Tang and P.C. Fife, Propagating Fronts for Competing Species Equations with Diffusion, Arch. Rat. Mech. Anal. 73 (1980), pp. 69-77.

28 A. Wunderlin and H. Haken, Scaling Theory for Non-equilibrium Systems, Z. Physik B 21 (1975), pp. 393-401.

29 H. Berestycki and P.L. Lions, Existence d'ondes solitaires dans des problèmes nonlinéaires du type Klein-Gordon, C.R. Acad. Sci. Paris, Ser. A, t. 288, pp. 395-398.

30 B. Nicolaenko, A General Class of Nonlinear Bifurcation Problems from a Point in the Essential Spectrum. Application to Shock Wave Solutions to Kinetic Equations, in : P.H. Rabinowitz (ed.), Applications of Bifurcation Theory, Academic Press 1977, pp. 333-357.

Michael RENARDY
Institut für theoretische Physik
Pfaffenwaldring 57,
7000 Stuttgart 80

SERGIO SPAGNOLO

Quelques remarques sur la G-convergence

Dans cet exposé, je me propose d'illustrer les résultats principaux de la G-convergence elliptique et quelques méthodes de démonstration.

D'un point de vue très général, étant donnés des problèmes aux limites bien posés $P_k (k \in N)$ et P, on pourrait appeler G-convergence de $\{P_k\}$ vers P l'éventuelle convergence faible (dans l'espace où il y a une estimation "a priori" pour les solutions) des opérateurs de Green correspondants. Le fait de considérer la convergence faible au lieu de la convergence forte, a pour conséquence que la G-convergence est par sa nature compacte et qu'on peut approcher dans la G-convergence un problème quelconque par des problèmes beaucoup plus simples.

D'autre part on vérifie que, dans beaucoup de cas importants, la G-convergence est de fait indépendante de la condition aux limites particulière qu'on a voulu considérer. On appelle cette propriété *localité* de la G-convergence. Grâce à elle, on peut parler de G-convergence aussi pour des problèmes aux limites où il n'y a pas unicité de la solution, ou bien de G-convergence d'équations différentielles *locales*.

Pour être plus précis, plaçons-nous dans le cas le plus simple, celui des opérateurs elliptiques $A : H^1_{loc}(\Omega) \to H^{-1}_{loc}(\Omega)$ (Ω = ouvert de $\mathbb{R}^n$) qui peuvent se représenter sous la forme

$$A = - \operatorname{div}(a(x)D) \qquad (1)$$

où $a(x)$ est une matrice $n \times n$, mesurable en $x \in \Omega$, telle que

$$\left.\begin{aligned} &\lambda|\xi|^2 \leq (a(x)\xi,\xi) \leq \Lambda|\xi|^2 \\ &|(a(x)\xi,\eta)| \leq M(a(x)\xi,\xi)^{1/2}(a(x)\eta,\eta)^{1/2} \end{aligned}\right\} \quad (2)$$

pour tout $x \in \Omega$ et $\xi, \eta \in \mathbb{R}^n$, où $0 < \lambda \leq \Lambda$ et $M \geq 1$

Désignons par $\mathcal{E}(\lambda,\Lambda,M;\Omega)$ la classe de ces opérateurs différentiels. Chaque élément A de cette classe opère avec continuité de $H^1_{loc}(\Omega)$ à valeurs dans $H^{-1}_{loc}(\Omega)$; de plus, pour tout $\mu > 0$ (et même pour $\mu = 0$ lorsque Ω est borné) $\mu I + A$ induit un isomorphisme de $H^1_o(\Omega)$ sur $H^{-1}(\Omega)$: on notera $(\mu I+A)^{-1}_\Omega$ l'inverse de cet isomorphisme.

DEFINITION 1. Soient $A_k (k \in \mathbb{N})$ et A dans la classe $\mathcal{E}(\lambda,\Lambda,M;\Omega)$. On dit que $\{A_k\} \overset{G}{\to} A$ sur Ω pour $k \to \infty$, si

$$\{(\mu I+A_k)^{-1}_\Omega f\} \to (\mu I+A)^{-1}_\Omega f \quad \text{dans} \quad L^2(\Omega)$$

pour tout $f \in H^{-1}(\Omega)$, et pour un certain $\mu > 0$. □

On vérifie sans peine que la G-convergence est induite par une métrique sur $\mathcal{E}(\lambda,\Lambda,M;\Omega)$.

Voici les deux résultats de compacité et de localité :

THEOREME 1. (compacité). Toute suite $\{A_k\} \subseteq \mathcal{E}(\lambda,\Lambda,M;\Omega)$ admet une sous-suite G-convergente sur Ω vers un élément de la classe $\mathcal{E}(\lambda,M^2\Lambda,M;\Omega)$. □

THEOREME 2. (localité). Une condition nécessaire et suffisante pour qu'une suite $\{A_k\} \subseteq \mathcal{E}(\lambda,\Lambda,M;\Omega)$ soit G-convergente sur Ω vers A est

$$\left.\begin{array}{l}\forall \{u_k\} \subset H^1_{loc}(\Omega), \quad \forall h(k) \to \infty , \\ \underline{\text{si}}\ \{(u_k, A_{h(k)}u_k)\} \to (u,f) \text{ dans } L^2_{loc}(\Omega)\times H^{-1}_{loc}(\Omega), \\ \underline{\text{alors}}\ \ Au = f\ \underline{\text{sur}}\ \Omega \end{array}\right\} \quad (3)$$

□

Remarques : 1) Dans le cas particulier $M = 1$, le théorème 1 affirme que la classe $\mathcal{E}(\lambda,\Lambda,1;\Omega)$ est G-compacte

2) Le théorème 2 montre que la propriété "$\{A_k\} \overset{G}{\to} A$ sur Ω" ne dépend en réalité que des opérateurs $A_k : H^1_{loc}(\Omega) \to H^{-1}_{loc}(\Omega)$. Le théorème montre aussi que, si $\{A_k\} \overset{G}{\to} A$ sur Ω, alors $\{A_k\} \overset{G}{\to} A$ sur toute partie ouverte de Ω; en outre si $\{\Omega_i\}$ est une partition ouverte de Ω et si $\{A_k\} \overset{G}{\to} A$ sur $\Omega_i, \forall i$, alors $\{A_k\} \overset{G}{\to} A$ sur Ω tout entier.

3) Une autre façon d'exprimer la condition (3) est de considérer la convergence de Kuratowski d'une suite de parties. Rappelons que, si $\{S_k\}$ est une suite de parties d'un espace métrique X, on note X-Max lim S_k l'ensemble des éléments $x \in X$ tels que tout voisinage de x rencontre chaque élément $S_{k'}$ d'une sous-suite extraite de $\{S_k\}$, X-Min lim S_k l'ensemble des x tels que tout voisinage de x rencontre S_k dès que k est assez grand. Si le Max lim coïncide avec le Min lim, alors on dit que $\{S_k\} \to S$ au sens de Kuratowski.

Cela dit, la condition (3) n'est autre chose que

$$(L^2_{loc}(\Omega) \times H^{-1}_{loc}(\Omega)) - \operatorname*{Max\,lim}_{k\to\infty} \Gamma(A_k) \subseteq \Gamma(A) \qquad (3')$$

où $\Gamma(A_k)$ et $\Gamma(A)$ sont les graphes de A_k et A dans $H^1_{loc}(\Omega) \times H^{-1}_{loc}(\Omega)$

4) Supposons que $\{A_k\} \overset{G}{\to} A$ sur Ω . On peut alors (en résolvant des problèmes de Dirichlet convenables) trouver, pour tout $u \in H^1_{loc}(\Omega)$, une

suite $\{u_k\} \subseteq H^1_{loc}(\Omega)$ telle que $\{(u_k, A_k u_k)\} \to (u, Au)$ dans $L^2_{loc}(\Omega) \times H^{-1}_{loc}(\Omega)$. En d'autres termes $\Gamma(A)$ est contenu dans le $(L^2_{loc}(\Omega) \times H^{-1}_{loc}(\Omega))$ - Min $\lim_{k \to \infty} \Gamma(A_k)$

En conclusion, $\{A_k\} \xrightarrow{G} A$ sur Ω si et seulement si

$$\Gamma(A) = (L^2_{loc}(\Omega) \times H^{-1}_{loc}(\Omega)) - \lim_{k \to \infty} \Gamma(A_k)$$

On peut d'ailleurs préciser le résultat ci-dessus, en montrant que si $\{A_k\} \xrightarrow{G} A$ sur Ω, alors pour tout $u \in H^1_{loc}(\Omega)$ tel que $Au \in L^2_{loc}(\Omega)$, il existe $\{u_k\} \subseteq H^1_{loc}(\Omega)$ telle que $\{(u_k, A_k u_k)\} \to (u, Au)$ dans $L^2_{loc}(\Omega) \times L^2_{loc}(\Omega)$. Donc, une autre condition nécessaire et suffisante pour que $\{A_k\} \xrightarrow{G} A$ sur Ω est que

$$\Gamma_{L^2_{loc}}(A) = (L^2_{loc}(\Omega) \times L^2_{loc}(\Omega)) - \lim_{k \to \infty} \Gamma_{L^2_{loc}}(A_k)$$

où $\Gamma_{L^2_{loc}} = \Gamma \cap (L^2_{loc}(\Omega) \times L^2_{loc}(\Omega))$

Les théorèmes 1 et 2 furent d'abord prouvés (voir [1] et [2], ou bien [3]) dans le cas particulier où les opérateurs A_k sont symétriques, et ensuite ils furent prouvés dans le cas général (voir [4] ou [5]).

Pour mieux comprendre la différence entre les deux cas, il est convenable d'introduire la classe $\mathcal{M}(\lambda, \Lambda, M; \Omega)$ de toutes les matrices $a(x)$, mesurables en $x \in \Omega$, qui vérifient les conditions (2).

L'application $a(x) \to -\operatorname{div}(a(x)D)$, définie sur $\mathcal{M}(\lambda, \Lambda, M; \Omega)$ à valeurs dans $\mathcal{E}(\lambda, \Lambda, M; \Omega)$ est (par définition) surjective, mais elle n'est injective que dans le cas unidimensionnel $(n = 1)$. En d'autres termes, un opérateur $A \in \mathcal{L}(H^1_{loc}(\Omega), H^{-1}_{loc}(\Omega))$ qui admet une représentation du type $\{(1),(2)\}$, admet de fait plusieurs de ces représentations. Mais si parmi ces représentations il y en *une* où la matrice $a(x)$ est symétrique, alors celle-ci est *l'unique* représentation symétrique de A. En outre tout opérateur $A \in \mathcal{E}(\lambda,\Lambda,M;\Omega)$ qui est symétrique (au sens que $\langle Au,v\rangle = \langle Av,u\rangle$, $\forall u,v$ dans $\mathcal{D}(\Omega)$) admet une, et donc une seule, représentation {(1), (2)} avec $a(x)$ matrice symétrique. Remarquons enfin que $a(x)$ est une matrice symétrique si et seulement si elle vérifie (2) avec $M = 1$.

En conclusion, $\mathcal{E}(\lambda,\Lambda,1;\Omega)$ coïncide avec la classe des opérateurs elliptiques symétriques, et l'application $a(x) \to -\mathrm{div}(a(x)D)$ induit une bijection de $\mathcal{M}(\lambda,\Lambda,1;\Omega)$ sur $\mathcal{E}(\lambda,\Lambda,1;\Omega)$. Donc dans la classe (G-compacte, grâce au théorème 1) $\mathcal{E}(\lambda,\Lambda,1;\Omega)$ la G-convergence peut être aussi interprétée comme une convergence sur les coefficients, tandis que cela n'est pas possible en général.

On peut tout de même définir sur $\mathcal{M}(\lambda,\Lambda,M;\Omega)$ une convergence, dite H-convergence (voir [5] ou [6]), qui entraîne la G-convergence des opérateurs associés.

<u>DEFINITION 2</u> : Soient $\{a_k(x)\}$ et $a(x)$ dans la classe de matrices $\mathcal{M}(\lambda,\Lambda,M;\Omega)$. On dira que $\{a_k(x)\} \overset{H}{\to} a(x)$ sur Ω pour $k \to \infty$, si (ayant notées $u_k(f) \equiv (\mu I - \mathrm{div}(a_k(x)D))_\Omega^{-1} f$ et $u(f) \equiv (\mu I - \mathrm{div}(a(x)D))_\Omega^{-1} f$ les solutions des problèmes de Dirichlet sur Ω) on a

$$u_k(f) \to u(f) \quad \text{dans} \quad L^2(\Omega)$$

et

$$\{a_k(x).Du_k(f)\} \to a(x).Du(f) \quad \text{dans } [L^2(\Omega)]^{n^2}\text{-faible},$$

pour tout $f \in H^{-1}(\Omega)$, et pour un $\mu > 0$. □

On prouve ([5] ou [6]) pour la H-convergence un résultat analogue au théorème 1 : toute suite $\{a_k(x)\}$ dans $\mathcal{M}(\lambda,\Lambda,M;\Omega)$ admet une sous-suite H-convergente dans Ω vers une matrice $a(x)\in\mathcal{M}(\lambda,M^2\Lambda,M;\Omega)$.

Puisque (comme on voit aussitôt en confrontant la définition 2 avec la définition 1) l'application $a(x) \to -\operatorname{div}(a(x)D)$ est continue de $\mathcal{M}(\lambda,\Lambda,M;\Omega)$ muni de la H-convergence sur $\mathcal{E}(\lambda,\Lambda,M;\Omega)$ muni de la G-convergence, ce résultat est plus fort que le théorème 1. On remarquera en outre que l'application ci-dessus induit un homéomorphisme entre les deux classes $\mathcal{M}(\lambda,\Lambda,1;\Omega)$ et $\mathcal{E}(\lambda,\Lambda,1;\Omega)$.

Remarque : On peut aussi interpréter la H-convergence comme une convergence au sens de Kuratowski.
Désignons par $P(a) : H^1_{loc}(\Omega) \to L^2_{loc}(\Omega) \times H^{-1}_{loc}(\Omega)$, où $a\in\mathcal{M}(\lambda,\Lambda,M;\Omega)$, l'opérateur qui fait correspondre à $u\in H^1_{loc}(\Omega)$ le couple $(a(x)Du, -\operatorname{div}(a(x)Du))$.
On remarquera que cet opérateur caractérise de façon univoque la matrice $a(x)$.
Or, on vérifie que $\{a_k(x)\} \overset{H}{\to} a(x)$ sur Ω si et seulement si

$$\Gamma(P(a)) = (L^2_{loc}(\Omega)\times H^{-1}_{loc}(\Omega)\times H^{-1}_{loc}(\Omega)) - \lim_{k\to\infty} \Gamma(P(a_k))$$

Pour conclure, je ferai quelques considérations relatives aux techniques de démonstration des théorèmes 1 et 2. Je me bornerai au cas particulier des opérateurs symétriques.

La démonstration des théorèmes 1 et 2 qui se trouve dans [1] ,[2] ou [3], repose sur les deux lemmes suivants (qui présentent un intérêt en eux-mêmes):

LEMME 1 : (voir [7]) Soit $A \in \mathcal{L}(H_0^1(\Omega), H^{-1}(\Omega))$ un opérateur tel que

$$\begin{cases} \langle Au, v\rangle = \langle Av, u\rangle \\ \lambda\int_\Omega |Du|^2\, dx \leq \langle Au, u\rangle \leq \Lambda \int_\Omega |Du|^2 dx \end{cases} \tag{4}$$

où $0 < \lambda \leq \Lambda$, $\forall u, v$ dans $H_0^1(\Omega)$, et tel que

$$\text{supp}(Au) \subseteq \text{supp}(Du), \qquad \forall u \in H_0^1(\Omega).$$

Alors A est différentiel, i.e. $A \in \mathcal{E}(\lambda, \Lambda, 1; \Omega)$ □

[J'ignore si un résultat analogue est valable dans le cas d'opérateurs non-symétriques].

LEMME 2 : (voir [1] ou [3]). Soit $\{A_k\} \subset \mathcal{E}(\lambda, \Lambda, 1; \Omega)$ et $A \in \mathcal{L}(H_0^1(\Omega), H^{-1}(\Omega))$ tels que

$$\{(\mu I + A_k)_\Omega^{-1} f\} \to (\mu I + A)_\Omega^{-1} f \quad \text{dans } L^2(\Omega) \tag{5}$$

pour tout $f \in H^{-1}(\Omega)$ et pour un $\mu > 0$.

Alors

$$\text{supp}(Au) \subseteq \text{supp}(Du), \quad \forall u \in H_0^1(\Omega) \tag{6}$$ □

Pour prouver le Lemme 2, on peut utiliser les opérateurs paraboliques engendrés par les A_k (comme dans [1]) ou bien se servir des approximants de Yosida (comme dans [3]). J'indique ici une troisième méthode, peut-être la plus naturelle du moment qu'il s'agit d'un problème de supports, qui utilise les équations hyperboliques.

Soit $A \in \mathcal{L}(H_0^1(\Omega), H^{-1}(\Omega))$ un opérateur qui vérifie (4). On peut alors considérer, pour tout $\phi \in H_0^1(\Omega)$, le problème de Cauchy

$$w'' + Aw = 0 \qquad \text{sur } \Omega \times [0,+\infty[$$

$$w(0) = \phi, \quad w'(0) = 0 \qquad \text{sur } \Omega$$

où $w(t) \in H_0^1(\Omega)$ pour tout $t \geq 0$.

On sait que ce problème admet une (et une seule) solution $w \equiv T_A(t)\phi$ dans $L^\infty([0,+\infty[,H_0^1(\Omega))$.
On vérifie en outre sans peine que

$$\frac{1}{t^2} \langle T_A(t)\phi - \phi, v\rangle \longrightarrow -\frac{1}{2}\langle A\phi, v\rangle, \text{ pour } t \to 0^+, \tag{7}$$

pour tout $v \in H_0^1(\Omega)$.

Enfin, l'intégrale

$$u = \frac{1}{\mu} \int_0^{+\infty} e^{-\mu t}\, T_A(t)\phi \;\, dt$$

est convergente dans $H_0^1(\Omega)$, $\forall \mu > 0$ et $\phi \in H_0^1(\Omega)$, et

$$\mu^2 u + Au = \phi \quad \text{sur } \Omega\ .$$

Grâce à cela, on prouve que si $\{A_k\}$ est dans $\mathcal{L}(H_0^1(\Omega),H^{-1}(\Omega))$ et "G-converge" vers A au sens de (5), alors on a, $\forall t \geq 0$

$$\{T_{A_k}(t)\phi\} \to T_A(t)\phi \quad \text{dans } L^2(\Omega), \quad \forall \phi \in H_0^1(\Omega)\ .$$

Soit maintenant $\{A_k\}$ une suite d'opérateurs différentiels (i.e. $\{A_k\} \subseteq \mathcal{E}(\lambda,\Lambda,1;\Omega)$) qui vérifie (5) : on se propose de prouver que A vérifie (6). Or (6) signifie que si ϕ est constante sur un ouvert $\Omega_0 \subseteq \Omega$, alors $A\phi \equiv 0$ sur Ω_0. Soit donc $\phi \equiv$ Cste sur un ouvert $\Omega_0 \subseteq \Omega$.

En utilisant le fait que A_k est un opérateur différentiel, on en déduit que la solution $T_{A_k}(t)\phi$ du problème de Cauchy est constante, et égale à ϕ, sur le *cône de dépendance*

$$\Gamma_{\Omega_o} = \{(x,t) : \text{dist}(x, C\Omega_o) > t\sqrt{\Lambda}\}$$

En particulier on a. $\forall v \in \mathcal{D}(\Omega_o)$:

$$\langle T_{A_k}(t)\,\phi - \phi, v\rangle = 0 \quad , \quad \forall t < \frac{1}{\sqrt{\Lambda}}\,\text{dist}(\text{supp}(v), C\Omega_o),$$

d'où, pour $k \to \infty$, on tire (grâce à (8)) que

$$\langle T_A(t)\;\phi - \phi, v\rangle = 0 \text{ , pour les mêmes } t.$$

De là on obtient, en faisant $t \to 0^+$ (cf. (7)) que

$$\langle A\phi, v\rangle = 0 \quad \forall v \in \mathcal{D}(\Omega_o) \text{ , i.e. que } A\phi = 0 \text{ sur } \Omega_o, \quad \text{C.Q.F.D.}$$

Bibliographie

1 S. Spagnolo, Sul limite delle soluzioni di problemi di Cauchy relativi all'equazione del calore. Ann. Scu. Norm. Pisa, 21 (1967), p. 657-699.

2 S. Spagnolo, Sulla convergenza di soluzioni di equazioni paraboliche ed ellittiche. Ann. Scu. Norm. Pisa, 22 (1968), p. 575-597.

3 S. Spagnolo, Convergence in energy for elliptic operators. Synspade 1975. Academic Press New York 1976, p. 469-498.

4 L. Tartar, Convergence d'opérateurs différentiels Analisi Convessa e Applicazioni, Roma 1974 (Quaderno C.N.R.), p. 101-104.

5 L. Tartar, *Cours Peccot* au Collège de France, Paris 1977.

6 F. Murat, *H-Convergence*. Séminaires de l'Université d'Alger (1978).

7 S. Spagnolo, *Una caratterizzazione degli operatori differentiali*.. Rend. Sem. Padova 39 (1967), p. 56-64.

Sergio SPAGNOLO

Scuola Normale Superiore
di Pisa
Piazza dei Cavalieri

PISA - ITALIE